Stop Faking It!

Finally Understanding Science So You Can Teach It

ELECTRICITY AND MAGNETISM

Stop Faking It!

Finally Understanding Science   So You Can Teach It

ELECTRICITY AND MAGNETISM

NATIONAL SCIENCE TEACHERS ASSOCIATION

Arlington, Virginia

NATIONAL SCIENCE TEACHERS ASSOCIATION

Claire Reinburg, Director
Judy Cusick, Senior Editor
Andrew Cocke, Associate Editor
Betty Smith, Associate Editor

ART AND DESIGN Linda Olliver, Director
  Brian Diskin, Illustrator
PRINTING AND PRODUCTION Catherine Lorrain, Director
  Nguyet Tran, Assistant Production Manager
  Jack Parker, Electronic Prepress Technician
New Products and Services, SCILINKS Tyson Brown, Director
  David Anderson, Database Web and Development Coordinator

NATIONAL SCIENCE TEACHERS ASSOCIATION
Gerald F. Wheeler, Executive Director
David Beacom, Publisher

09 08 07 5 4 3

Library of Congress Cataloging-in-Publication Data
Robertson, William C.
  Electricity and magnetism : stop faking it! finally understanding science so you can teach it/by
William C. Robertson.
    p. cm.
  Includes index.
  ISBN 0-87355-236-9
  1. Science—Study and teaching—Popular works. 2. Science—Methodology—Popular works. I.
Title.
  Q181.R568 2004
  537'.07'1—dc22
                    2004013339

NSTA is committed to publishing quality materials that promote the best in inquiry-based science education. However, conditions of actual use may vary and the safety procedures and practices described in this book are intended to serve only as a guide. Additional precautionary measures may be required. NSTA and the author(s) do not warrant or represent that the procedure and practices in this book meet any safety code or standard or federal, state, or local regulations. NSTA and the author(s) disclaim any liability for personal injury or damage to property arising out of or relating to the use of this book including any recommendations, instructions, or materials contained therein.

Permission is granted in advance for photocopying brief excerpts for one-time use in a classroom or workshop. Requests involving electronic reproduction should be directed to Permissions/NSTA Press, 1840 Wilson Blvd., Arlington, VA 22201-3000; fax 703-526-9754. Permissions requests for coursepacks, textbooks, and other commercial uses should be directed to Copyright Clearance Center, 222 Rosewood Dr., Danvers, MA 01923; fax 978-646-8600; *www.copyright.com.*

SCILINKS
THE WORLD'S A CLICK AWAY *Featuring SciLinks®—a way to connect text and the Internet. Up-to-the-minute online content, classroom ideas, and other materials are just a click away. Go to page x to learn more about this educational resource.*

Contents

Preface

The book you have in your hands is the fifth in the *Stop Faking It!* series. The previous four books have been well received, mainly because they stick to the principles outlined below. All across the country, teachers, parents, and home-schoolers are faced with helping other people understand subjects—science and math—that they don't really understand themselves. When I speak of understanding, I'm not talking about what rules and formulas to apply when, but rather knowing the meaning behind all the rules, formulas, and procedures. I *know* that it is possible for science and math to make sense at a *deep level*—deep enough that you can teach it to others with confidence and comfort.

Why do science and math have such a bad reputation as being so difficult? What makes them so difficult to understand? Well, my contention is that science and math are *not* difficult to understand. It's just that from kindergarten through graduate school, we present the material *way* too fast and at too abstract a level. To truly understand science and math, you need *time* to wrap your mind around the concepts. However, very little science and math instruction allows that necessary time. Unless you have the knack for understanding abstract ideas in a quick presentation, you can quickly fall behind as the material flies over your head. Unfortunately, the solution many people use to keep from falling behind is to *memorize* the material. Memorizing your way through the material is a surefire way to feel uncomfortable when it comes time to teach the material to others. You have a difficult time answering questions that aren't stated explicitly in the textbook, you feel inadequate, and let's face it—it just isn't any fun!

So, how do you go about *understanding* science and math? You could pick up a high school or college science textbook and do your best to plow through the ideas, but that can get discouraging quickly. You could plunk down a few bucks and take an introductory college course, but you might be smack in the middle of a too-much-material-too-fast situation. Chances are, also, that the undergraduate credit you would earn wouldn't do the tiniest thing to help you on your teaching pay scale. Elementary and middle school textbooks generally include brief explanations of the concepts, but the emphasis is definitely on the word *brief*, and the number of errors in those explanations is higher than it should be.

Finally, you can pick up one or fifty "resource" books that contain many cool classroom activities but also include too brief, sometimes incorrect, and vocabulary-laden explanations.

Given the above situation, I decided to write a series of books that would solve many of these problems. Each book covers a relatively small area of science, and the presentation is slow, coherent, and hopefully funny in at least a few places. Typically, I spend a chapter or two covering material that might take up a paragraph or a page in a standard science book. My hope is that people will take it slow and digest, rather than memorize, the material.

This fifth book in the series is about electricity, magnetism, and electrical circuits. To best understand the material, it's best to have lots of semi-expensive material to play around with. Assuming most of us (writers and readers!) don't have the big bucks to spend, we've come up with a great alternative, which is special software that allows one to simulate simple to complex electrical circuits on the computer (see page xi for details). Aside from the fact that this software allows us to build circuits that would otherwise cost lots of money, we have the added advantage that electrocution is a remote possibility at best. On top of those values, you get a genuinely cool piece of software that is just plain a fun toy.

There is an established method for helping people learn concepts, and that method is known as the **learning cycle**. Basically, it consists of having someone do a hands-on activity or two, or even just think about various questions or situations, followed by explanations based on those activities. By connecting new concepts to existing ideas, activities, or experiences, people tend to develop understanding rather than memorization. Each chapter in this book, then, is broken up into two kinds of sections. One kind of section is titled, "Things to do before you read the science stuff," and the other is titled, "The science stuff." If you actually do the things I ask prior to reading the science, I guarantee you'll have a more satisfying experience and a better chance of grasping the material.

It is important that you realize this book is *not* a textbook. It is, however, designed to help you "get " science at a level you never thought possible, and also to bring you to the point where tackling more traditional science resources won't be a terrifying, lump-in-your-throat, I-don't-think-I'll-survive experience.

Dedications

This being the fifth book in this series, I feel a set of dedications are long overdue. I first dedicate this book to Marie Galpin, who taught me that it was okay to be different and to pursue my own goals. You always accepted me for who I was when I doubted that. Thank you, Marie.

Second, I would like to thank David Beacom, Claire Reinburg, and the rest of the kind and efficient group at NSTA Press for taking a chance on this series. They trusted that teachers would find the series entertaining and that it would turn out to be one of those rare commodities—a fun medium for learning.

I would like to thank Brian Diskin for proving the old adage wrong—you *can*, in fact, tell a book by its cover. As I surreptitiously watch people browse through the titles at the NSTA conventions, they rarely pass by the *Stop Faking It!* books without at least picking them up and thumbing through the pages. The humor and overall style is apparent from the book covers and from the obvious quality of the page-by-page illustrations. Brian's art entices the reader to dig just far enough to see what else the book has to offer.

Finally, I would like to thank my wife, Jann, and my children, Sara and Jesse, for putting up with the fact that Dad has to work way too many nights and weekends in order to finish these silly books. All three also serve as my best critics.

Acknowledgments

The *Stop Faking It!* series of books is produced by the NSTA Press: Claire Reinburg, director; Andrew Cocke, project editor; Linda Olliver, art director; Catherine Lorrain-Hale, production director. Linda Olliver designed the cover from an illustration provided by artist Brian Diskin, who also created the inside illustrations.

This book was reviewed by Daryl Taylor (Williamstown High School, New Jersey); Kenneth Thompson (Emporia State University, Kansas); and Larry Kirkpatrick (Montana State University).

About the Author

As the author of NSTA Press's *Stop Faking It!* series, Bill Robertson believes science can be both accessible and fun—if it's presented so that people can readily understand it. Robertson is a science education writer, reviews and edits science materials, and frequently conducts inservice teacher workshops as well as seminars at NSTA conventions. He has also taught college physics and developed K–12 science curricula, teacher materials, and award-winning science kits. He earned a master's degree in physics from the University of Illinois and a PhD in science education from the University of Colorado.

About the Illustrator

The soon-to-be-out-of-debt humorous illustrator Brian Diskin grew up outside of Chicago. He graduated from Northern Illinois University with a degree in commercial illustration, after which he taught himself cartooning. His art has appeared in many books, including *The Golfer's Personal Trainer* and *5 Lines: Limericks on Ice*. You can also find his art in newspapers, on greeting cards, on T-shirts, and on refrigerators. At any given time he can be found teaching watercolors and cartooning, and hopefully working on his ever-expanding series of *Stop Faking It!* books. You can view his work at *www.briandiskin.com*.

How can you avoid searching hundreds of science Web sites to locate the best sources of information on a given topic? SciLinks, created and maintained by the National Science Teachers Association (NSTA), has the answer.

In a SciLinked text, such as this one, you'll find a logo and keyword near a concept, a URL (*www.scilinks.org*), and a keyword code. Simply go to the SciLinks Web site, type in the code, and receive an annotated listing of as many as 15 Web pages—all of which have gone through an extensive review process conducted by a team of science educators. SciLinks is your best source of pertinent, trustworthy Internet links on subjects from astronomy to zoology.

Need more information? Take a tour—*www.scilinks.org/tour*

Instructions for Downloading Virtual Labs: Electricity

For PC users (Windows)

(1) Go to *ftp://ftp.nsta.org/VLabsWin/*

(2) Double click on the file *Vlabs_Electricity_Setup.exe*

(3) When the prompt comes up, click "Save."

(4) Choose the location where you want to save the file.

For MAC users

(1) Go to *ftp://ftp.nsta.org/VLabsMac/*

(2) Double click on the file *Vlabs_Electricty_DL.bin*

(3) When the prompt comes up, click "Save."

(4) Choose the location where you want to save the file.

Virtual Labs: Electricity is a product of Sunburst Technology. All rights reserved.

Small Sparks to Get Us Going

J udging from the title of this chapter, you might guess that I'm starting off this book with a few basics about electricity, and you'd be right. Other than that great guess of yours, you might be wondering why the book is about electricity *and* magnetism. Is there any special reason to put these two areas of science together? The answer is yes. In fact, the connection between electricity and magnetism is so strong that you can actually view them as the same thing. You'll have to wait until Chapter 4, though, to see that connection. Sorta gets you all anxious and nervous waiting for that, huh? In the meantime, let's get moving on those electricity basics.

Actually, I'm going to warn you about something first. For many of the activities I'll be asking you to do in this and in the next chapter, it's best if you do them when the humidity isn't very high. What that means is that these activities will work best in the winter (when it's generally not as humid), and will work best if you live in places like Colorado, Arizona, Nevada, and California. I'm not saying these activities *won't* work where it's really humid, but rather that you won't get dramatic results. Of course, if you live someplace like Houston or Tampa, it's the middle of August, and the skies are threatening rain, maybe you should put this book away and try another day.

Things to do before you read the science stuff

Grab a balloon, blow it up, and tie it off. Look at yourself in a mirror as you rub the balloon vigorously on your hair and then slowly pull the balloon away from your head. You should get a result something like that shown in Figure 1.1. By the way, this won't work if your hair is wet or if you use hair spray or gel on your hair. Also, don't even think about getting any results if you have a buzz cut or dreadlocks or a shaved head (duh!). Notice that if you pull the balloon a long ways from your hair, your hair relaxes back down. Bring the balloon slowly toward your hair and there's an attraction again.

Figure 1.1

Now that you have this fun trick, why not share it with a friend? Get that friend and make sure he or she has hair that satisfies the previous requirements (not wet, no gel, etc.). Rub the balloon on *your* hair and then hold the balloon near your friend's hair. Not much happening? Try holding the balloon near *your* hair to convince you the trick still works. Just for closure, have your friend rub the balloon on his or her hair and then place the balloon near each of the two heads in the room. What happens?

Blow up a second balloon and tie it off. Rub both of your balloons on your hair, or on a carpet if you don't have the right kind of hair for this thing. With the balloons resting in the palms of your hands, slowly bring the balloons together. When they get very close together, you should notice something happening, such as the balloons pushing each other away.

Finally, shuffle your feet across a carpeted room (this works best if you're wearing thick socks or tennis shoes) and then touch something metal. If the air is dry enough, you'll get a nice shock. If it's too humid, just think back to the last time you got a shock from touching a doorknob or a drinking fountain. Of course if you're not crazy about giving yourself a shock, you can just rub your balloon on the carpet and then hold it near a metal doorknob. Listen carefully and you'll hear a spark jump between the balloon and the doorknob. Darken the room and you can *see* the spark jump. Cool.

The science stuff

A long time ago, in a galaxy far, far away...er...in a country in Europe, early Greek philosophers studied things like electric shocks that you sometimes get from touching metal objects and the apparent attraction that some materials (such as amber[1] and hair) have for each other after you rub them together. Then around the year 1700, scientists were so interested in these occurrences that they set up a bunch of controlled experiments in which they studied what happens when various materials come in contact with one another, resulting in attractions or even sparks. To explain what was going on, these scientists made up[2] the model that the world contains positive charges and negative charges.[3] Part of this model is the fact that positive charges and negative charges like to be together, so that when we separate them, they have a tendency to jump back together. In other words, there's an **attractive force**[4] between positive and negative charges.

Armed with this simple model, we can explain what happened with the balloon and your hair. Because positive and negative charges are attracted to each other, most things in the world have an equal number of positive and negative charges. Something with an excess of positive charges attracts negative charges until the numbers are equal. As such, the normal state of the world is

---

[1] Amber is a resin that is a bit like rubber or plastic. The Greek word for amber is *electron*, a bit of info that explains why we call all of this stuff electricity.

[2] If the words "made up" bother you, allow me to explain. People invent science concepts and models, which we can refer to collectively as scientific explanations. These concepts and models are not written in stone somewhere waiting to be discovered. Of course, even though people make up these ideas, there are agreed-upon rules by which one judges scientific models. Good models have to explain all sorts of observations and even predict new observations. Good scientific explanations stick around and bad ones die away. Suffice to say that the concepts and models I use in this book have been around a while.

[3] The names positive and negative are just names. We could just as easily call them red charges and blue charges. It turns out that the plus and minus designation is useful, though, when considering the directions of electric forces.

[4] *Force* is the name scientists give to any push, pull, hit, nudge, etc. For more than you ever wanted to know about forces, see the *Stop Faking It!* book on Force and Motion.

Figure 1.2

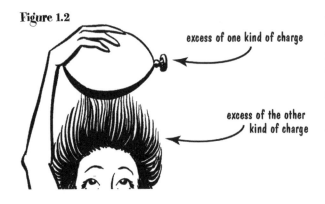

excess of one kind of charge

excess of the other kind of charge

that objects are **electrically neutral** (equal numbers of positive and negative charges). Now evidently, when you rub two materials together, it's possible for either positive or negative charges to move from one material to the other, resulting in one material having an excess positive charge and the other an excess negative charge. In this situation, shown in Figure 1.2, one object (your hair, for instance) is attracted to another object (the balloon).

In addition to experiencing this attractive electric force, you also saw two charged balloons *repel* each other. Why? Well, each balloon picked up the same kind of charge (either positive or negative) when you rubbed them on your hair. Apparently this means that like charges repel. So, we have maybe the one thing you remember from studying electricity in elementary school, which is that

Opposite charges attract and like charges repel.

You also should have noticed that electric forces get stronger as the objects get closer together and weaker as the objects get farther apart. The two balloons don't push on each other appreciably until you get them close together, and the charged balloon stops pulling strongly on your hair when you get the balloon far enough away from your head. The fact that electric forces decrease with distance is really important for explaining what you'll observe in the next section. One other important fact that we shouldn't just gloss over is that once separated, the two kinds of charges tend to get back together with each other (remember, opposite charges attract) so that the overall tendency of objects is to be electrically neutral. Charged balloons don't stay charged forever.

Let's take just a minute and reflect on this simple model of electric forces. We claim that objects contain both positive and negative charges, and also that opposite charges attract one another and like charges repel one another. The forces get weaker with increased separation of the charges. We have *made up* the existence of these charges to explain our observations. Of course, we haven't yet decided whether, when you rub a balloon on your hair, the balloon gets a net positive charge or a net negative charge. That's because the choice of which charges are positive and which are negative is *totally arbitrary*! Fortunately, we don't have to stew over which to call positive and which negative, because Ben Franklin already made the arbitrary choice for us.

We can update that simple model of positive and negative charges moving around and exerting forces on one another. Currently scientists use a model stating that material consists of things called atoms. Maybe you've heard of them. An **atom** is composed of a central nucleus, which contains positively charged protons and neutral neutrons (clever name, huh?), surrounded by a rather vague cloud of negatively charged electrons.[5] See the drawing in Figure 1.3, and notice that I did not draw an atom like a solar system, with electrons in orbits around the nucleus. Turns out that model, while useful for explaining a few things, just doesn't cut it in the long run. Also, the thing keeping electrons near nuclei (plural of **nucleus**) is that simple model that opposite charges attract. Nuclei, containing positive protons, attract negative electrons.

Topic: The Electron

Go to: *www.scilinks.org*

Code: SFEM01

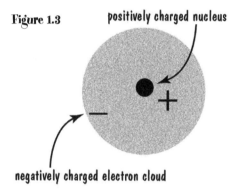

Figure 1.3

positively charged nucleus

negatively charged electron cloud

Anyway, it turns out that different materials have atoms with different numbers of **protons**, **neutrons**, and **electrons**. Also, some nuclei hang onto their electrons stronger than other nuclei. When you put different materials together, sometimes the strong-attracting nuclei steal electrons from the weaker-attracting nuclei. That makes electrons jump from one material to another. In this way, the atoms in a balloon steal electrons from the atoms in your hair.[6] That gives the balloon an excess negative charge and your hair an excess positive charge, which causes your hair to move toward the balloon.

Okay, what's going on when you rub a balloon on your head and then place it near a friend's hair? Little to no attraction, right? The secret here is that rubbing the balloon on your hair left your hair positively charged and the balloon negatively charged. Big attraction. Your friend's hair is neutral, because you didn't steal any electrons from it. Although a negatively charged object *can*

---

[5] The question I hope pops into your mind is, "How in the world do we know that atoms exist, and if they do, how do we know that they're composed of protons, neutrons, and electrons?" Great question, but I'm afraid I can't answer it within the scope of this book. Unfortunately, you'll just have to accept that this is a scientific model that works, and that there are lots of experiments that make atoms like these a good explanation of how the world works. Of course, atoms still are just a model. You don't have to believe in them if you don't want to!

---

[6] Technically, the atoms in both the balloon and your hair are part of larger things called *molecules*, and it is correct to say that the *molecules* steal electrons from each other. For now, I'll just stick with the atom description and not complicate matters. The process is the same for individual atoms or complex molecules.

attract a neutral object (see the next section), that attraction generally isn't nearly as strong as the attraction between a negatively charged object and a positively charged object.

Finally, what about shuffling your feet across a carpet and then touching something metal? Pretty simple, actually. When you shuffle your feet on the carpet, you steal electrons from the carpet. You now have an excess of electrons, and a resulting negative charge. When you get near the metal object, those excess electrons jump over to the metal (I'll explain why in the next "science stuff" section). You can see and feel the resulting spark when these electrons jump.

We refer to the business of charges exerting forces on other charges as **static electricity**. Static means not moving, and in all the examples we talked about so far, the charges are stationary once they go from one object to another. Now it might seem silly to refer to this as static electricity when the charges do, in fact, move from one place to another. It turns out, though, that there is another phenomenon known as **current electricity**, in which the charges are in constant motion. So, when the charges are stationary for a fair amount of time, we call it static electricity.

SCI**LINKS**
THE WORLD'S A CLICK AWAY

Topic: Static Electricity

Go to: *www.scilinks.org*

Code: SFEM02

Topic: Current Electricity

Go to: *www.scilinks.org*

Code: SFEM03

More things to do before you read more science stuff

Still have one of those inflated balloons? If not, get another one. Tear a piece of paper into tiny bits and place them on a table. Next rub the balloon on your hair, on a carpet, or on a mellow dog. This will give the balloon excess electrons and an overall negative charge, right? Now slowly bring the charged balloon near the bits of paper. Do your best to stop before the bits of paper actually jump up to the balloon. They should do a nice little dance for you. For a change of pace, you can substitute a pile of pepper for the tiny bits of paper. You're going to do a few more fun things with balloons in this section, but before you do them, see if you can answer the following question: If the bits of paper and the pepper are electrically neutral (which they are), how come they're attracted to a negatively charged balloon?

Okay, on to more fun things. Place a Ping-Pong ball on a smooth surface. Charge up your balloon again (hair, carpet, dog) and bring it near the Ping-Pong ball without touching it. Be sure to hold the balloon slightly above and to the side of the Ping-Pong ball, as shown in Figure 1.4. I guess this means Ping-Pong balls like balloons. Repeat this using an empty aluminum pop can in place of the

Ping-Pong ball. Again hold the balloon slightly above and to the side of the can.

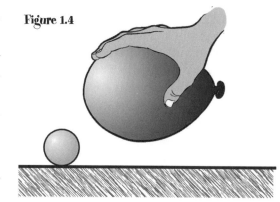

Figure 1.4

Get some bubble solution and blow a few bubbles. Bring the charged balloon near one of the floating bubbles, again without touching the bubble with the balloon. This proves that bubbles like balloons, too! With a little practice, you can use a charged balloon to make a bubble rise and fall repeatedly, sort of like you have a string attached to the bubble, as our magician friend is doing in Figure 1.5.

Figure 1.5

balloon

soap bubble

Final magic trick before you get to the Applications section. Charge up your balloon again and bring it near a thin stream of water. You should be able to deflect the stream of water to the side as long as you don't touch the stream with the balloon (Figure 1.6).

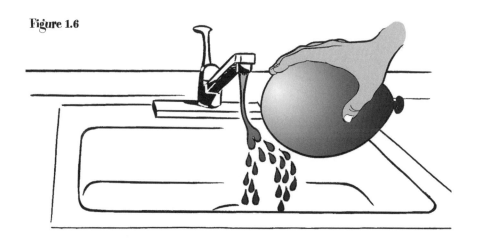

Figure 1.6

More science stuff

In every activity in the previous section, you brought a charged object (the balloon) near an uncharged, or neutral, object or substance (the paper bits, the Ping-Pong ball, the bubbles, the water, and so on). These all resulted in attractive forces. Let's see how that might happen. For starters, Figure 1.7 shows a simplified model of what the atoms in a tiny piece of paper look like. This is a simplified model because the atoms in a piece of paper are of many different kinds, and they are organized into molecules. Figure 1.7 just shows orderly rows of atoms, each with a positively charged nucleus and a negatively charged cloud of electrons around the nucleus.

Figure 1.7

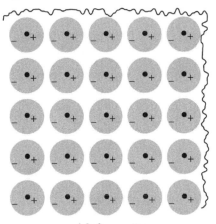

model of paper atoms

In this model, the positively charged nuclei are essentially stationary, and the **electron clouds**, while not free to move all over the place, are able to move a little bit.

Okay, what happens when you bring a negatively charged balloon near the paper? Because like charges repel and unlike charges attract, the balloon pushes the electron clouds away and attracts the nuclei. Only the electron clouds, however, are able to move a little bit. What happens, then, is that each atom in the paper gets slightly distorted, as in Figure 1.8.

Figure 1.8

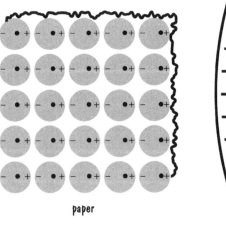

paper

balloon

While these slight distortions don't change the overall structure of the paper, they do create a situation where, *on average,* the positively charged nuclei are closer to the balloon than the negatively charged electron clouds, as in Figure 1.9.[7] At this point, you have to recall that electric forces get weaker as distance increases. That means that, *on average,* the repulsion between the electron clouds and the balloon (like charges repel) is *weaker* than the attraction between the nuclei and the balloon (unlike charges attract).

Figure 1.9

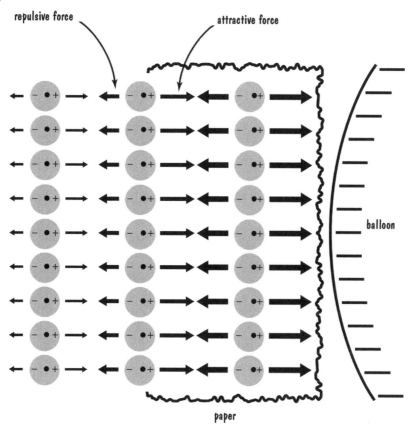

With the attraction being stronger than the repulsion, the balloon and the paper attract each other. The paper jumps up to the balloon. In this way, a charged object can attract a neutral object. Scientists say that the negatively

---

[7] In some resource materials and textbooks, you will read that the electrons in the paper, being pushed away from the balloon, rush to the other side of the piece of paper. That just ain't so, because the electrons in paper do not have that much freedom of movement; they can't jump from one atom to another.

charged balloon has **induced** a charge in the paper, and the paper is said to contain an **induced charge**. That term sorta makes sense, because the paper doesn't really have an overall net charge, but the presence of the balloon makes it acts like it does. Kind of like *induced labor*, which isn't naturally occurring labor but sure as heck has the same result!

All the other things you did in the previous section rely on the balloon inducing a charge in a neutral object. The pepper is exactly like the bits of paper, except there you would talk about the atoms in the pepper instead of the atoms in the paper. The Ping-Pong ball is also just about the same procedure, but I really ought to explain the necessity of holding the balloon in the proper place in order to get the Ping-Pong ball to roll. As with the paper and the pepper, the negatively charged balloon induces a charge in the Ping-Pong ball, resulting in a net attractive force (check out Figure 1.10).

Figure 1.10

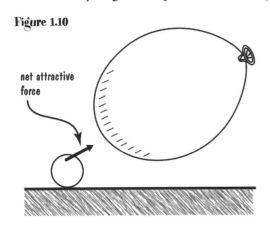

net attractive force

When you apply a force as shown in Figure 1.10, you will cause the ball to roll. Contrast this with what would happen if you held the balloon directly to the side of the Ping-Pong ball. In that case, you'd exert a force on the ball such that you would make it *slide* instead of roll (see Figure 1.11). It's easier to roll a Ping-Pong ball than slide it (you have to work against friction in the latter case), and the electric force between the balloon and the ball generally isn't strong enough to slide the ball.

Figure 1.11

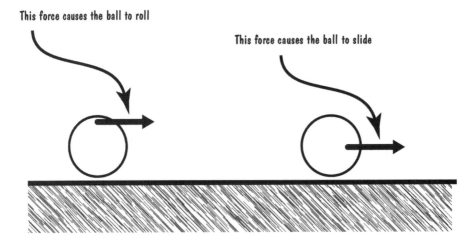

This force causes the ball to roll

This force causes the ball to slide

On to the aluminum pop can. The position of the balloon again is important (you want to roll the can rather than slide it), but there is a big difference in what the atoms in the can are doing. The current scientific model for what's inside metals (aluminum is a metal!) is a bunch of positively charged nuclei that don't move around, just as with nonmetals like paper, but with electrons that are basically free to move around inside the metal.[8] Therefore, when you bring a negatively charged balloon near the metal, all the electrons in the metal tend to move away from the negative balloon (like charges repel). You end up with something like Figure 1.12.

Figure 1.12

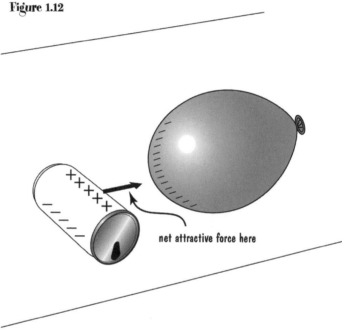

net attractive force here

The net effect is the same. Because electric forces get weaker with distance, the attraction of the balloon to the positive side of the can is stronger than the repulsion of the balloon for the negative side of the can. The overall attraction for the can is larger than the overall attraction for the Ping-Pong ball (you'll notice that it's easier to use the balloon to roll the can than to roll the Ping-Pong ball) because the charge separation is greater when the electrons are free to move around (the metal can) than when all you can do is distort the individual atoms a bit (the Ping-Pong ball).

The last things to explain are the soap bubbles and the stream of water. They're basically the same process, so I'll just explain the stream of water. The scientific model for what's inside liquids is a bit different from that for solids. The model for solids is that the nuclei are essentially motionless (the nuclei actually move a bit, but not enough for us to worry about), and the model for liquids is a collection of molecules that kind of slip and slide over, under, and

---

[8] The actual accepted model for metals is somewhat more complicated than electrons just running around all over the place. Only some of the electrons in a metal are really free to roam, with mild restrictions on their motion. For our purposes here, however, we'll just stick with the picture of stationary nuclei and electrons that have a free run of the metal.

Figure 1.13

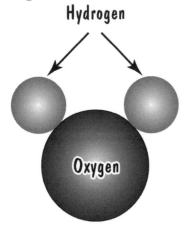

across one another. Water molecules contain one oxygen atom and two hydrogen atoms, arranged so they look just a bit like that mouse that lives in Florida and California (see Figure 1.13).

The electrons in a water molecule are shared by all three atoms, but they tend to spend more time around the oxygen atom than they do around the hydrogen atoms. This leaves one end of the molecule positively charged and the other end negatively charged. Such an arrangement, by the way, is known as an **electric dipole**, where "di-" means "two." Thus, a dipole has "two poles," one positive and one negative, shown in Figure 1.14.

Figure 1.14

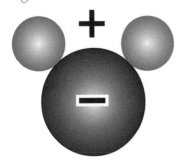

Figure 1.15

stream of water

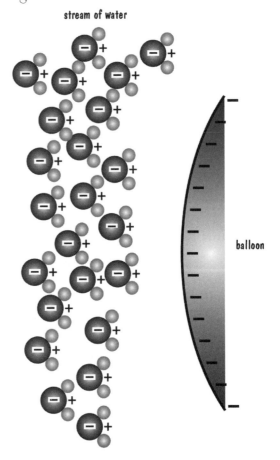

balloon

Let's look at a stream of water. It's a collection of lots and lots of these little hydrogen-oxygen dipoles. The charged balloon exerts a significant force only on the water molecules closest to it (remember, electric forces decrease with increased distance). Because the negatively charged balloon *attracts* the positive end of the water molecules and *repels* the negative end of the water molecules, the molecules closest to the balloon rotate so the oxygen end is away from the balloon and the hydrogen end is close to the balloon (see Figure 1.15).

Just as with our bits of paper, the negative charges are, *on average,* farther from the balloon than the positive charges. This results in a net attraction to the balloon, and the stream of water deflects. The difference between the water and the paper is that in one case you get a charge separation by the rotation of molecules and in the other you get a charge separation by distortion of individual atoms, as shown in Figure 1.16.

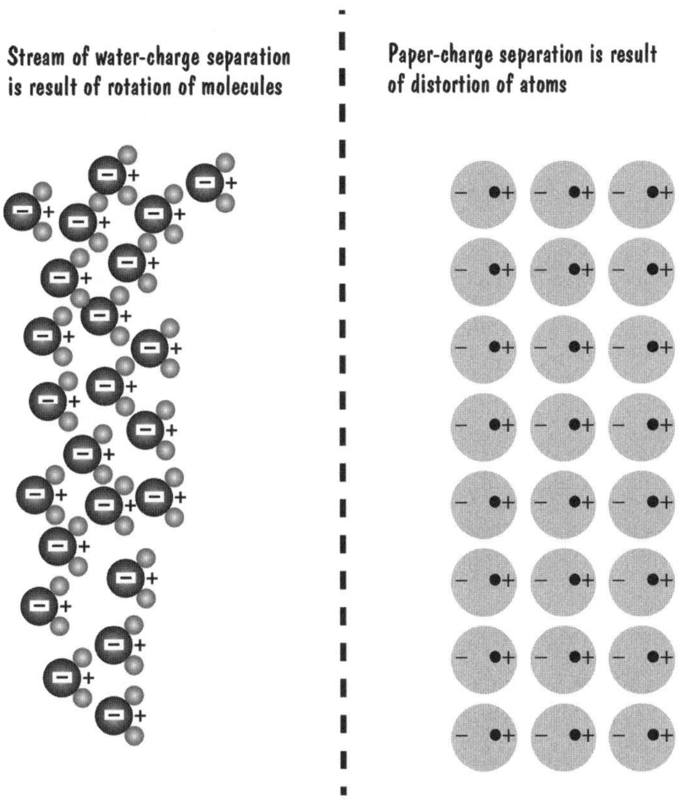

The attraction of a charged balloon for soap bubbles is just about the same as for the water, what with bubbles and water both being liquids. The only difference is that, in addition to having water molecules involved in the reorientation, bubbles have other molecules, such as soap and maybe glycerin, involved.

Figure 1.16

Stream of water-charge separation is result of rotation of molecules

Paper-charge separation is result of distortion of atoms

Oops, I lied. The water stream wasn't the last thing to explain in this section. I promised earlier that I'd explain why electrons jump from your hand to a metal object after you shuffle your feet on a carpet. Simple enough. When you bring your negatively charged finger near the metal object, the negatively charged electrons in the metal flee from that finger.[9] That leaves a positively charged region in the metal near your finger. That positive charge attracts the excess electrons you have with you to the point that they jump across. The actual jumping of electrons is an example of current electricity. More on that in later chapters.

---

[9] As the Monty Python folks would say, "Run away ...... run away!"

Chapter Summary

- There are two kinds of electric charges in the world—positive and negative.

- Static electricity deals with the electric forces between stationary charges.

- Like charges repel and unlike charges attract.

- Atoms in substances are composed of negatively charged electrons surrounding a positively charged nucleus.

- Electric forces get smaller as the distance between charges increases.

- You can *induce* a charge in a substance by creating a separation between the positive and negative charges in the atoms or molecules in the substance.

- An object that is positive on one end and negative on the opposite end is called an electric dipole.

- A large number of the electrons in metals are essentially free to move throughout the metal.

Applications

1. When I was a kid, trucks carrying flammable liquids used to have a chain attached to the truck that dragged on the ground. Every once in a while, you can still see trucks dragging chains. That chain isn't a show of force, but an attempt at safety. What follows is the incorrect thinking that people used to have leading to attaching chains to trucks. When cars and trucks move along the highway, the tires tend to pick up electrons from the pavement and transfer them to the body of the car or truck. That can lead to the buildup of lots of excess electrons. Too many electrons, and you can get a big ol' spark between the vehicle and the road, similar to the spark that jumps when you touch a metal doorknob or drinking fountain. Big ol' sparks can be a really bad thing if you have a load of flammable stuff. So, the chain attached to those trucks provides a path for the excess electrons to return to the road, minimizing the chance of any sparks. Unfortunately, that's not what happens, which is why attaching chains to trucks doesn't do what it's supposed to do. Tires do indeed pick up excess electrons from the road, but they don't transfer those electrons to the body of the truck. Instead, they simply drive the electrons in the truck away from the negatively charged tires. If you have a chain attached, the negative charge on the tires actually drives electrons off the truck and onto the road, leaving the truck with an overall positive charge. This actually *increases* the chance of a spark between the truck body and the tires or between the truck body and the road. So, the next time you see a truck with a chain dragging from it, politely go up and

explain to the driver the silliness of having that chain. Okay, maybe you shouldn't do that. What you *might* do is watch carefully the next time you're in an airplane that is being refueled. The process of adding fuel can transfer excess electrons to the plane, so the people doing the fueling attach a wire from the plane to the tarmac. This wire allows the plane to remain neutral (excess electrons head to the tarmac), eliminating the chance of a spark that could cause problems.

2. Maybe as a kid you rubbed balloons on your clothes or your hair and then placed them on a wall, where they would stick. Why do they stick? Simple. The charged balloon creates an induced charge in the wall, and there's an attraction. The balloon stays charged up for a while because the excess electrons on the balloon don't readily jump over to the wall, unless the wall is made of metal, in which case this trick won't work!

3. Right about now, I'm sure you're saying to yourself, "Balloons and hair and Ping-Pong balls are fine, but what about that annoying static cling?" Again, pretty simple. All that tumbling around in a dryer causes some clothes, such as those made of nylon, to pick up extra electrons from the other clothes. These clothes stay charged up if the weather's pretty dry, and they cling to other clothes and to you by inducing charges in other substances. Of course, static cling can be a good thing. In the manufacture of plastic food wrap and in the process of removing plastic wrap from the roll, the plastic wrap picks up and retains extra electrons. When you stretch a sheet of plastic wrap over a bowl, it sticks to the sides of the bowl because the negative charge on the plastic wrap induces a charge in the bowl, leading to an electric attraction.

4. Here's a good way to amuse yourself at bedtime. I know there are other ways, but sometimes those aren't available! On a relatively dry day (not humid), turn the lights out and then lift the covers or sheet repeatedly. In between the covers you'll see a bunch of little sparks. These are nothing more than the result of electrons transferring from one cover or sheet to another and then jumping back.

5. If you live where the humidity is low, this might be the best application of science you ever read. If it's a day when you're getting shocks every time you walk a bit and then touch something, try carrying your car keys in your hand. Before you touch a doorknob or drinking fountain or whatever, hold a key out and touch the object with that. This will discharge you (rid you of excess electrons) so you don't get a shock when you then touch the object with your hand. Of course, a spark does jump between the object and the key, and electrons do eventually go from your hand to the key to the object. Why you don't get a shock when that happens has to do with the amount of surface area involved in the charge transfer, and the fact that there really

isn't much of a gap across which the electrons jump. The more surface area involved, the less the shock. Holding a key involves a lot more surface area than extending your finger toward an object. All this means that if you don't have a key, you can significantly reduce any potential shock by touching things with your fist (larger surface area) rather than your finger. And with all this advice, we really ought to contact Heloise.

More about Charging Things

I n the first chapter, you charged things up with excess electrons and saw what could happen as a result. Now we're going to take things a step further and get an idea of how to measure how much charge something has, figure out whether an object is positively or negatively charged, and make up a new concept called the electric field. Hmmm ... something new to make up in each chapter!

Things to do before you read the science stuff

Gather together a clear drinking glass, a small plastic lid (such as comes with a margarine tub) that completely covers the top of the drinking glass, a paper clip, a small nail or pushpin, and some aluminum foil. Oh yeah, you'll again be needing that all-important piece of static electricity equipment—a balloon. Use the nail or pushpin to make a small hole in the center of the plastic lid.

Figure 2.1

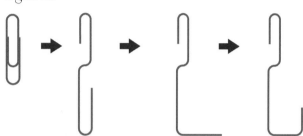

Next bend the paper clip as shown in Figure 2.1. You'll probably need a pair of pliers to make the small bend in the lower end of the clip.

Thread the bent paper clip through the hole in the lid and crimp some aluminum foil into a ball at the top of the paper clip, so it looks like Figure 2.2.

When you now put the lid on the glass, the flat bottom part of the paper clip should be at least 3 or 4 centimeters from the bottom of the glass, and it shouldn't touch the sides. You might have to use a bit of tape where the paper clip enters the lid in order to keep the paper clip in place. Cut a strip of aluminum foil that's about half a centimeter wide and 6 or 7 centimeters long. Use your fingers to smooth out and flatten the strip on a smooth surface, and then crease the strip in the middle. Cut the strip in two equal pieces, and then tape those pieces back together without a gap in-between. This might seem a silly step given that you cut the strip and then put it back together, but the tape makes everything work better. Lay the joined aluminum strip over the flat part of the paper clip. When you put the whole contraption together, it

Figure 2.2

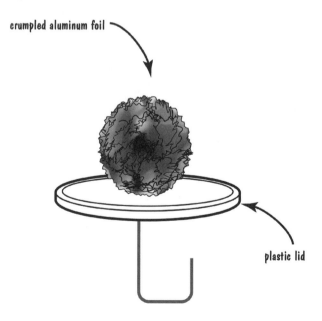

crumpled aluminum foil

plastic lid

should look something like Figure 2.3.

What you've just made is known as an **electroscope**. The strip of aluminum foil is called the leaves of the electroscope (each side is one leaf). Time to find out what in the world this device is good for. Charge up your balloon by rubbing it on your clothes or something similar. Slowly bring the balloon near the foil ball on top of the electroscope, approaching it from directly over the electroscope and not from the side (see Figure 2.4).

As you do this, watch the leaves. They should separate more and more as the balloon gets nearer the top of the electroscope. If both leaves move to one side or another, that means you're not approaching the electroscope from directly above. If you get too close to the electroscope, a spark will jump from the balloon to the aluminum foil ball on the top. What happens to the leaves when the spark jumps? What happens when you take the balloon away? With the balloon removed, touch the top of the electroscope with your finger. The leaves should return to their original position. If they don't, try wetting your finger a bit and touching the electroscope again.

The science stuff

There's not much new going on here. The aluminum foil top, the paper clip, and the aluminum foil leaves are all touching and act like one piece of metal, with electrons free to roam all around in the metal. When you bring a balloon with excess electrons near the ball of aluminum foil on top, the electrons in the metal are repelled (like charges repel) and move away from the ball. Where do they go? Why, as far away from

Figure 2.3

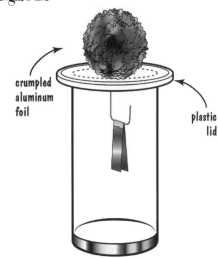

crumpled
aluminum
foil

plastic
lid

Figure 2.4

Approach from
directly above

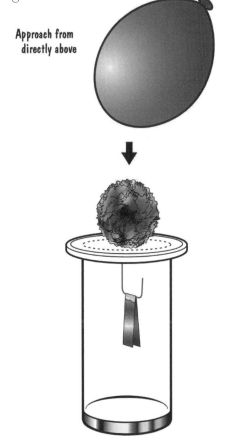

Figure 2.5

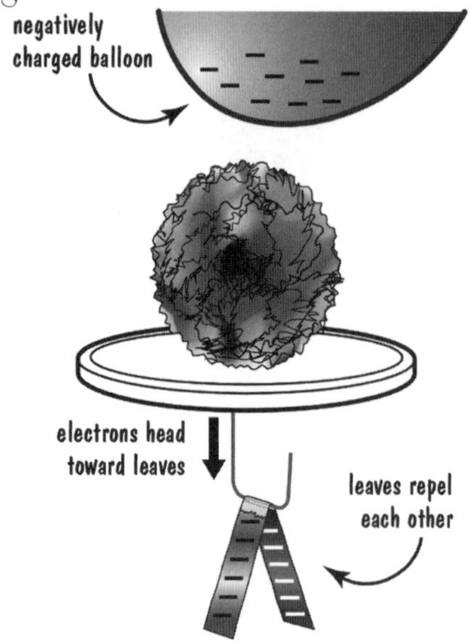

negatively
charged balloon

electrons head
toward leaves

leaves repel
each other

that balloon as they can. The leaves of the electroscope now have an excess of electrons, and each leaf is negatively charged. The negatively charged leaves repel each other, as in Figure 2.5.

You should also notice that the closer the balloon, the farther apart the leaves move. Remember that electric forces get stronger as you get closer, so the closer the balloon to the top of the electroscope the stronger the repulsive force on the electrons in the metal. This leads to more excess electrons in the leaves and a stronger repulsive force between the leaves. Hence, they move farther apart.

When you move the balloon away from the electroscope, the leaves collapse, right? That's because with the balloon removed, the electrons in the metal spread out to their original configuration, and all parts of the electroscope are neutral again. Ah, but what if you get the balloon so close that a spark jumps to the electroscope? In that case, the electroscope now has an overall negative charge. Even when you take the balloon away, there are still excess electrons all over the electroscope, including the leaves. Therefore, the leaves still repel each other (Figure 2.6). This process is known as charging the electroscope by conduction.

Figure 2.6

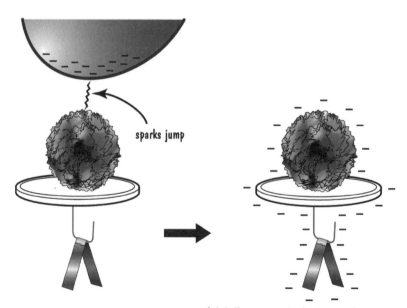

sparks jump

With balloon removed, electroscope has excess
electrons and remains negatively charged

Now, like all things, the electroscope has a tendency to be electrically neutral. Those excess electrons would leave the electroscope if you gave them a chance. By touching your finger to the top of the electroscope, you provide an escape route for the electrons. Those excess electrons travel through your finger to the rest of your body and on into the building you're in and eventually to the Earth itself (Figure 2.7). This process is known as **grounding**.

Topic: Conductors/
      Nonconductors

Go to: *www.scilinks.org*

Code: SFEM04

Figure 2.7

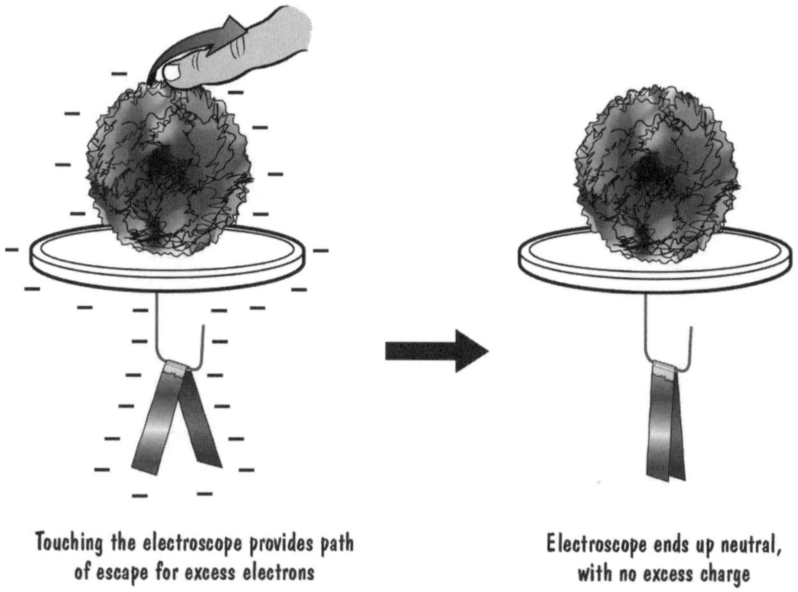

Touching the electroscope provides path
of escape for excess electrons

Electroscope ends up neutral,
with no excess charge

The Earth as a whole is electrically neutral, and it's so large that it serves, essentially, as an endless source of electrons and an endless reservoir into which electrons can flow. So, once you touch the electroscope, the electroscope becomes neutral, and you're ready to start over again with the balloon.

You undoubtedly noticed that if you brought the balloon in toward the electroscope from the side, or even just not directly down from the top, both leaves moved toward the balloon. This is an example of what we observed in Chapter 1, where the balloon exerts an overall attractive force on the aluminum foil leaves by inducing a charge in them. It's just like what happens with the balloon and aluminum can.

Before we move on, notice that if you have a really sensitive electroscope and a way to measure the separation of the leaves, you have a device that will tell you

how strong an electric force is present at the top. The stronger the force (in our case caused by the balloon getting closer), the greater the separation of the leaves. Expensive electroscopes use leaves made of gold foil, with the whole thing encased in glass. Such an electroscope can be calibrated to measure the relative strengths of different electric forces. I know—you can hardly contain yourself!

More things to do before you read more science stuff

Time for more fun with electroscopes. First ground the electroscope by touching it. Then charge up your balloon and bring it near enough to the electroscope so the leaves separate, but not so near that a spark jumps. If you get too close and a spark jumps, just ground the electroscope and start over. Now, with the charged balloon near and the leaves separated, touch the top of the electroscope with a finger of your free hand. Remove the finger and then pull the balloon away (the order of operations here is important!). Follow the steps shown in Figure 2.8.

The leaves of the electroscope should remain separated after you remove the balloon. Next bring the balloon near the electroscope again, but not so near that a spark jumps. Watch the leaves carefully as you do this. They should collapse at first, and then separate again. What the heck is going on here?

Figure 2.8

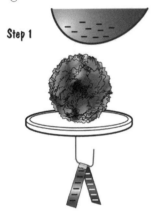

Step 1

Bring balloon close to electroscope

Step 2

Touch foil while balloon is near

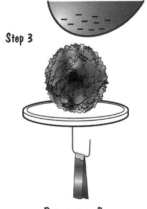

Step 3

Remove your finger

Step 4

Pull the balloon away

Topic: Electromagnetic
Induction

Go to: *www.scilinks.org*

Code: SFEM06

More science stuff

What you just did is known as charging the electroscope by **induction.** By bringing the negatively charged balloon near the electroscope, you got a bunch of electrons in the metal so scared they ran away to the leaves, just as before. Now, though, you touch the electroscope with your finger. Oh my, this is an even better escape route! The electrons can now head into your finger and get even farther from the balloon than if they went to the leaves. Notice that the leaves collapse as you touch the electroscope (Figure 2.9).

Okay, you have now driven electrons off of the electroscope, into your finger, and on into the ground.[1] That leaves the electroscope with an overall positive charge because it lost some negative charges. By removing your finger from the electroscope with the balloon still near, you leave the electroscope with an overall positive charge. You can see that the electroscope is charged because the leaves of the electroscope remain apart when you remove the balloon. Now the leaves are repelling each other not because they both have a negative charge, but because both have a positive charge (Figure 2.10).

So now you have an electroscope with an overall positive charge. Why do the leaves collapse when you bring a negatively charged balloon near the top

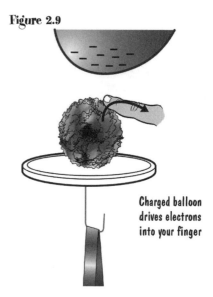

Figure 2.9

Charged balloon drives electrons into your finger

Figure 2.10

Electroscope is left with a positive charge because the balloon drove the electrons onto your finger

---

[1] I really should say something about the path the electrons take. It's not as if one group of electrons follows the entire path from the electroscope to your finger to your body and on into the ground. Electrons right next to your finger do in fact jump over to your finger, but that just causes other electrons to move a bit, which causes other electrons to move a bit. Thus, the *effect* of moving a small number of electrons gets transmitted along the entire path. The electrons that move from your feet onto the ground are certainly not the same ones that jumped from the electroscope to your finger, but the net effect is the same, so we'll continue to talk about electrons traversing the entire path.

of the electroscope? Because the presence of the balloon repels electrons, driving them toward the leaves. If you add electrons to the leaves, they get to the point that the negative charge on the electrons there balances the positive charge on the nuclei. Being neutral, the leaves collapse. Of course, although the leaves are neutral, the entire electroscope still has a net positive charge. As you bring the balloon even closer to the electroscope, you drive even more electrons into the leaves, making the leaves negatively charged overall. They repel. Slowly remove the balloon and you'll see the leaves collapse again and then repel as the balloon is completely removed (steps shown in Figure 2.11).

Lest you get too confused, let's recap what we've done with an electroscope. You can charge an electroscope by conduction just by bringing a charged object

Figure 2.11

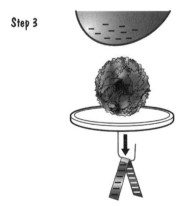

Step 1

Positvely charged electroscope

Step 2

Charged balloon drives electrons into leaves, making them neutral

Step 3

As balloon gets closer, it drives more electrons into leaves, giving them a negative charge

near enough to the electroscope that a spark jumps, or if a spark doesn't jump, by actually touching the ball of the electroscope with the charged object. If the object you bring near the electroscope has a negative charge, as with a rubbed balloon, then touching the electroscope will charge it negatively. If the object you bring near the electroscope has a positive charge (we haven't experienced that), then touching the electroscope will charge it positively.[2] You also can charge an electroscope by induction. To do that, you bring a charged object near the electroscope and then ground the electroscope by touching it. If the object is charged negatively, this grounding will give the electroscope a positive charge. If the object is charged positively, this grounding will give the electroscope a negative charge.[3] As part of this recap, take a moment and think about the overall silliness of our description! We're talking about made-up things called electrons that are so tiny no one, not even the residents of Whoville, can ever see them. These made-up electrons carry with them the made-up concept of negative charge, and this can make large objects either positively or negatively charged overall. Ain't science great? Of course, the remarkable thing is that our made-up scientific model does an excellent job of explaining what we observe.

Even more things to do before you read even more science stuff

Keep your electroscope handy while you collect the following materials: A Styrofoam tray that comes with meat you buy at the grocery store,[4] a small aluminum pie tin, a pair of scissors, and some tape. If the Styrofoam tray is dirty, clean it first. Then cut an L-shaped piece from the corner of the tray so you have a small Styrofoam hockey stick (see Figure 2.12).

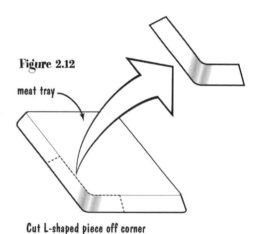

Figure 2.12

meat tray

Cut L-shaped piece off corner

---

[2] The reason this charges the electroscope positively is that electrons from the electroscope jump across to the positively charged object. When you remove negatively charged electrons from the electroscope, that leaves it with a positive charge.

---

[3] When you ground the electroscope in this latter situation, electrons move from your finger to the electroscope instead of from the electroscope to your finger.

---

[4] Tell the butcher you're a poor, starving teacher and the butcher might just give you a new tray that you don't have to clean.

Figure 2.13

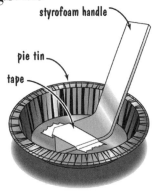

styrofoam handle

pie tin

tape

Tape your hockey stick to the inside bottom of the pie tin as shown in Figure 2.13, so the Styrofoam piece serves as a handle with which you can pick up the pie tin.

Next charge up the bottom of the rest of the Styrofoam tray by either rubbing it on your hair or rubbing it with a cloth. Place the tray bottom-side up on a flat surface. Then place the pie tin on top of the tray. Touch the pie tin with your finger (this might give you a small shock). Then use the Styrofoam handle to lift the pie tin off the tray. Touch the pie tin again and you should get another shock (Figure 2.14). Repeat what you've just done a bunch of times to prove to yourself that each time you go through the procedure you can charge up the pie tin. Where's all this charge coming from? Does the Styrofoam tray have an infinite supply of electrons or something?

Get your electroscope and use a charged balloon to charge up the electroscope negatively. In other words, bring the balloon close enough to the electroscope that a spark jumps. Next place your pie tin again on top of the charged-up Styrofoam tray. Touch the pie tin and then remove it using the handle. This time do *not* touch the pie tin again. Instead, bring the pie tin near the ball of the electroscope from directly above the electroscope. Watch what happens to the leaves, and then see if you can figure out whether the pie tin has a positive charge or a negative charge. Oh boy, a mystery!

Figure 2.14

Step 1
Touch pie tin while it's on the meat tray

charged meat tray

Step 2
Touch pie tin again after you remove it from the meat tray

Even more science stuff

The device you just built is known as an **electrophorus.** Scientists used to use a device like this as an easy way to charge things up over and over without having to ruin their coiffures. Here's how an electrophorus works. First, you give the tray an excess of electrons by rubbing it with a cloth or by rubbing it on your hair. It turns out that Styrofoam is a substance that *really* likes electrons, so much so that when you give it extra electrons, it holds onto those extra electrons so tightly that it's really difficult to remove the electrons from the Styrofoam. So, when you place the pie tin on top of the tray, virtually no electrons jump over to the pie tin. Something does happen in the pie tin, though. The electrons in the pie tin are repelled by the negative charge on the Styrofoam tray, and they move away from the tray. When you touch the pie tin, you are grounding it and providing a path for those electrons to move even farther away from the tray. That leaves the pie tin with an overall...all together now...positive charge. You pick up the pie tin using the handle, and you have a positively charged pie tin. If you now touch the pie tin, electrons in your finger are *attracted* to the pie tin and jump across, making the pie tin neutral. Figure 2.15 illustrates this whole procedure.

So, is the Styrofoam tray an infinite source of extra charge? Not at all. The pie tin gets charged up because *you* are accepting or donating electrons to the pie tin. All the tray is doing is holding on to the extra electrons you gave it in the beginning.

By now you should be so good at this business of electrons moving around that you can explain why the negatively charged electroscope did what it did when you brought the charged pie tin near. Okay, okay, I'll tell you. First, I hope you no-

Figure 2.15

Step 1
Electrons in pie tin "escape" to your finger

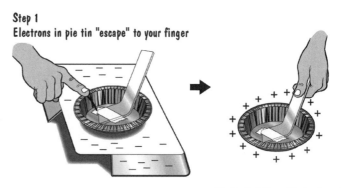

Pie tin is left positively charged

Step 2
Electrons from your finger move toward the positively charged pie tin

Pie tin is left neutral

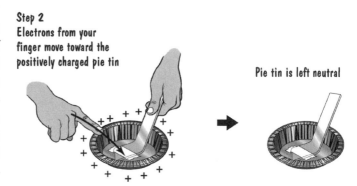

ticed the following. When you charge the electroscope negatively, the leaves remain separated. When you bring the pie tin near, the leaves initially collapse. That's because the pie tin is positively charged. It attracts electrons away from the leaves, causing them to collapse because they become neutral. If you bring the pie tin close enough, you can attract enough electrons away from the leaves that they become positively charged, and they repel again. And whether you know it or not, we have now discovered that electroscopes are good for two things. By measuring the separation of the leaves, you can use them to tell how strong an electric force is being exerted. By charging them up either positively or negatively, you also can use them to determine whether another charged object is positive or negative. Bringing a positively charged object near a negatively charged electroscope causes the leaves to collapse. Bringing a negatively charged object near a negatively charged electroscope causes the leaves to move farther apart.

And even more things to do before you read even more science stuff

This next thing I'm going to have you do is kind of gross, but it's also pretty neat, so let's hope the neat part outdoes the gross part. Get an unused clear bottle of baby oil and remove the label. Then get a small lock of hair. You can use human hair (yours or someone else's—just wait until they're asleep), dog hair, or even doll hair (Barbie does science!). Use scissors to cut this hair into lots of tiny pieces, and I do mean tiny. Cut across the hair in one direction and then rotate these pieces ninety degrees and cut again. Keep cutting until you have a pile of itty-bitty bits of hair. Now for the gross part. Open the bottle of baby oil and put all the hair bits into the baby oil. Put the lid back on the bottle and shake vigorously until the hair bits are distributed more or less evenly throughout the oil.

Get your trusty balloon and charge it up. Then bring the balloon near the bottle of hair and oil. Watch what the hairs do as you bring the balloon close and then remove it. By the way, this is the neat part of the activity. Just in case you're having trouble seeing the hairs, they should look something like Figure 2.16.

Figure 2.16

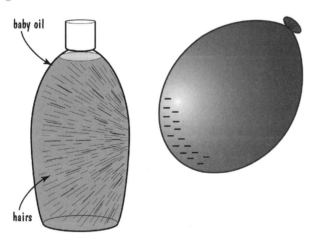

baby oil

hairs

And even more science stuff

Before I explain why those hairs did what they did, and what that might mean, I'm going to give you a formula (yes, I said a formula—don't panic!) that you can use to figure out just how big a force one charged object exerts on another. We're not going to use that formula here, but I want you to see it and know what it means so you don't get freaked out if you come across it in a textbook. Anyway, here's the relationship expressed in words:

Electric force between two objects = (a number) $\dfrac{\textit{(charge of one object)(charge of other object)}}{\textit{(distance between the objects)}^2}$

Because this is the first formula in this book, let me explain a few basics about algebraic expressions. Whenever two sets of parentheses are next to each other, that means you multiply the quantities inside the parentheses by each other. In other words, (122)(15) means multiply 122 and 15 together. The superscript 2 in the above formula means to square what's inside the parentheses right before it. And of course, that long line means divide what's above the line by what's below the line.

Now that we know that, let's see if this formula makes sense. The number on the right is just a number, so we can ignore it for now. It does make sense that the charge on *both* objects should be involved, because if either object has a larger or smaller charge, that should make the force correspondingly larger or smaller. Now on to the distance between the charged objects. We know that electric forces get weaker as the separation distance increases. Does the formula fit with that experience? Sure. The distance between the objects is in the *denominator* of the formula. If you increase that distance, does the force get smaller? To answer that, let's look at specific fractions. Suppose you have the fraction 1/2. What happens to the value of this fraction if we increase the denominator from 2 to 8? We get the fraction 1/8.   1/8 is a smaller number than 1/2, so increasing the denominator of a fraction gives you a smaller number overall. Therefore, in our formula for the electric force between two charged objects, increasing the distance between the objects increases the size of the denominator, and decreases the size of the force. In fact, because it's not just the distance but the distance *squared*, even a small increase in distance means the electric force gets quite a bit weaker.

Just so we can look all official and everything, I'm going to give you this formula as it looks in textbooks.

$$F_{electric} = k \frac{q_1 q_2}{r^2}$$

where k equals that number, which equals 9×10^9, which equals $9{,}000{,}000{,}000$[5]

q_1 = charge on one object

q_2 = charge on the other object

r^2 = distance between the objects[6]

This relationship is known as **Coulomb's law**, named for the French physicist Charles Coulomb. Not surprisingly, electric charge is measured in units of coulombs. It's interesting that even though the early Greek philosophers were aware of static electricity and its effects, it took until the 1700s for somebody like Coulomb to come up with an exact relationship for the magnitude of electric forces. Why'd it take so long, you ask? Most likely because electric interactions are kind of strange. When you push on something, you can see what's causing the force (you) and you can figure out relatively easily how to measure how big the force is just by how hard you push. With electric forces, you get things attracting and repelling one another, and you see and feel sparks, but it's not at all obvious what's causing all of that. Also, because electric forces are so strong, the tendency for objects to exchange charges and become neutral is great. That can really mess up experiments where you're trying to study electric forces.

This seems like a good opportunity to point out the difference between the book you have in your hands (this one) and regular textbooks. Most texts *begin* a discussion of electrostatic forces by introducing Coulomb's law. Those texts state that this formula describes how big the forces between charges are, and they proceed from there to solve various textbook-like problems. As with the development of ideas in all the books in the *Stop Faking It!* series, I have spent quite a bit of time using activities and explanations to lead you up to the concept—in this case a statement of Coulomb's law—in the hope that this abstract formula will make more sense when you have enough background *for* it to make sense.

---

[5] This number has units associated with it, just as the other symbols in this relationship have units. Because we don't spend a whole lot of time calculating things in this book, I didn't include the units above. For those of you dying to know, the units of k are $\dfrac{Newton \cdot Meters^2}{Coulombs^2}$

---

[6] This formula technically only works when the objects are infinitesimally small, as with electrons, or when the objects are spherical and r represents the distance between their centers. Since we're not going to do much in the way of calculating forces here, no need to worry about this technicality.

Okay, so we've got a formula that tells us how large an electric force two charged objects will exert on each other. How in the world do two things that aren't even touching each other provide a push or a pull? The answer is that we don't know! They just do. Of course, that doesn't stop us from creating a scientific model of what's going on. That model consists of the concept of a "field" that is caused by the presence of positive and negative charges and spreads out into space. Not surprisingly, it's known as an **electric field**, and we represent it with electric field lines. The electric field lines that surround a negatively charged balloon are shown in Figure 2.17.[7]

According to this model, some other object "sees" these field lines and responds accordingly. This is where all those tiny hairs suspended in the baby oil come in. Each hair gets an induced charge (negative charges being, on average, farther from the balloon and positive charges being, on average, closer to the balloon) due to the presence of the charged balloon, and "lines up," with negatives away from the balloon and positives close to the balloon. With our model of electric field lines, the explanation becomes even simpler. The hairs with induced charges simply line up with the electric field lines created by the charged balloon, as shown in Figure 2.18.

Compare the pattern you observed with the tiny hairs with the pattern of elec-

Figure 2.17

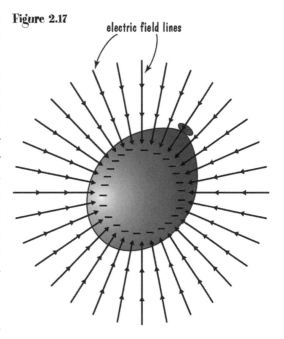

electric field lines

Figure 2.18

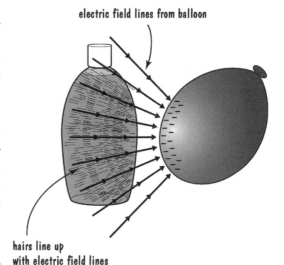

electric field lines from balloon

hairs line up with electric field lines

---

[7] This drawing assumes that the negative charge is spread out evenly across the balloon, which is not always the case. In fact, you might have noticed that the negative charge on the balloons you've been using is confined to one part of the balloon, as using the side of the balloon opposite the part you rubbed doesn't lead to significant electric forces.

tric field lines shown in Figure 2.17. In a way, the hairs are "revealing" the electric field lines to us.

To sum up this section, you can use Coulomb's law to figure out just how large a force one charged object will exert on another. If you draw the electric field lines created by a charged object, you can figure out what another charged object will do by figuring out its response to the electric field of the first object. And one more thing to add: There is a direction to electric field lines. They are drawn so positive charges feel a force in the direction of the field lines and negative charges feel a force opposite to the direction of the field lines, shown in Figure 2.19.

Figure 2.19

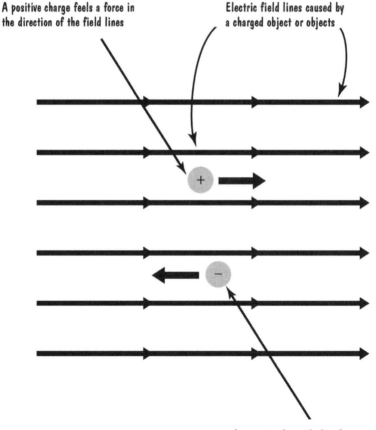

A positive charge feels a force in the direction of the field lines

Electric field lines caused by a charged object or objects

A negative charge feels a force opposite to the direction of the field lines

Chapter Summary

- An electroscope is a device we can use to measure the strength of the electric force exerted by a charged object as well as to determine whether the object is positively or negatively charged.

- You can charge an electroscope by *conduction* or *induction*.

- *Grounding* an electroscope or any other device means providing a path between the object and the Earth along which electrons can travel freely.

- An *electrophorus* provides a method for charging objects positively over and over.

- The mathematical expression that describes the electric force between two objects is known as Coulomb's law. This expression is written as

$$F_{electric} = k\frac{q_1 q_2}{r^2}$$

- Electric interactions can also be described using the concept of an *electric field*. Charged objects in the presence of the electric field lines created by another charged object feel a force. Positive charges feel a force in the direction of the field lines and negative charges feel a force opposite to the direction of the field lines.

Applications

1. You might be wondering just how much charge is jumping from one place to another when you charge up balloons and such. It turns out that not much charge is moving. Richard Feynman, a deceased, brilliant physicist and educator, put it this way.[8] If you were standing at arm's length from someone and each of you had just one percent more electrons than protons, the electrical repulsion would be strong enough to lift a "weight" equal to the weight of the entire Earth. This demonstrates two things. First, if an excess of electrons of one percent can create that large a force, then in our activities we're dealing with a whole lot less than one percent going from one place to another. Second, electric forces are *really* strong! Even tiny imbalances of charge can lead to large forces, which explains why the universe has a strong tendency toward neutral objects. Tiny imbalances lead to large forces, which tend to restore things to neutrality.

---

[8] This example is from the book series, *The Feynman Lectures on Physics*. (Feynman, Leighton, and Sands, 1964, Addison-Wesley.)

2. You can use an electroscope to figure out how much certain substances tend to hold onto or lose electrons. First charge up an electroscope positively using induction.[9] With the electroscope charged positively, the leaves will collapse in the presence of a negatively charged object and will repel farther in the presence of a positively charged object. Thus, you can use the electroscope to figure out whether an object is positively or negatively charged. Let's test this. Take a crumpled sheet of newspaper and rub it on a Teflon surface (pots and pans are a good source of Teflon!). Use your positively charged electroscope to see whether the newspaper ended up with a positive or a negative charge. You should discover that the newspaper ends up with a positive charge. This means that Teflon seems to like electrons more than paper. People have tested lots of different materials this way, and have come up with something known as the **Triboelectric series**. This is a list of materials categorized by how much they tend to take electrons away from other materials. Materials at the top of the list tend to give up electrons to materials lower down on the list. Below is a partial listing of the materials in the Triboelectric series.

- Human hands
- Asbestos
- Animal fur
- Glass
- Human hair
- Nylon
- Wool
- Lead
- Aluminum
- Paper
- Cotton
- Wood
- Hard rubber
- Gold
- Polyester

---

[9] In case you forgot how to do this, go back to the second "things to do" section in this chapter (page 22).

- Styrofoam
- Saran Wrap
- Vinyl
- Silicon
- Teflon

I can see at least 10 great jokes hiding in the process of rubbing each of the above materials on one another, but I guess I'll have to restrain myself.

3. Our model of metals is that they have electrons that are free to roam all over the place. When you place a charged object near a piece of metal, do the electrons just move and move forever? No, they just move until the forces that result from the separation of charges inside the metal result in forces that counteract the force exerted by the external charged object, as shown in Figure 2.20.

Figure 2.20

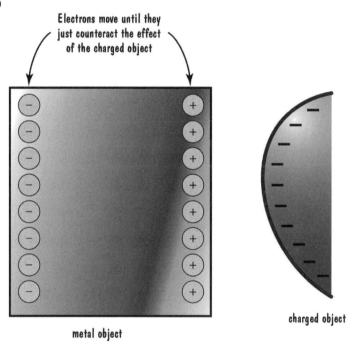

Electrons move until they just counteract the effect of the charged object

metal object

charged object

In terms of electric fields, what this means is that the charges in the metal move around until they completely cancel out the electric field caused by the external object. In other words, it's impossible to have a long-lasting electric field inside a piece of metal, or even a metal box. In fact even a metal cage, with holes in it, will do a good job of canceling external electric fields.

"So what?" you ask. Well, it turns out that radio waves are composed of electric fields and magnetic fields (I'll explain magnetic fields in the next chapter). If you are inside a metal cage, you have a difficult time receiving radio signals because the electrons in the metal cage move to cancel out the electric fields in the radio waves. Therefore, it's not a good idea to sit inside a metal cage if you want to listen to the radio. The only problem here is that we spend a lot of our lives inside metal cages (literally, not figuratively!). Large commercial buildings usually use metal frame construction, which is basically nothing more than a big metal cage. Hence, it's difficult for radio waves to penetrate inside commercial buildings. No radio reception and a tough time sending and receiving signals from cell phones.

4. If you've been paying attention, you know that atoms are composed of positively charged nuclei surrounded by negatively charged electrons. You also know that electric forces get stronger the closer the charges are. Why then, don't atoms just plain collapse as the nuclei attract the electrons to them? A good question that doesn't have an easy answer. The current theory, or scientific model, that applies to very small things like atoms is known as *quantum mechanics*. One of the consequences of that model is that it is just plain impossible for electrons to get too close to one another, which would have to happen if atoms collapsed. In other words, all our scientific models combined have to manage to explain what we observe and what we don't observe. Another question: If the nucleus of an atom is composed of positively charged protons and neutral neutrons, how come the nucleus doesn't just blow apart from the electric repulsion? After all, if protons are close together, then the value of r in Coulomb's law is extremely small, leading to a very large electric force. The answer is that there is an additional force, known as the *strong force*, which is much stronger than electric forces and holds the nucleus together. Again, two different scientific models combine to explain things. One final question: If electrons are negatively charged, how come individual electrons don't just blow apart from the electric repulsion? The answer here is we just don't know! As far as scientists know, electrons do not have any size at all—they are "point" objects. If they don't have any size, then they can't have separate parts to blow apart. And if that sounds unsatisfactory to you, you're not alone. Whether or not electrons have any size, and if so how to measure that size, is a long-standing question in physics. Keep in mind that scientific models are made up; they are not handed down from on high, etched on tablets. It's quite normal for scientific models to break down at some point and not be able to explain all things.

Magnets Enter the Picture

I started this book with the basics of electricity, and now it's time to introduce the basics of magnetism. What better way to do that than to play around with magnets? I promised earlier that I would explain how closely tied electricity and magnetism are, but that will have to wait until the next chapter. For now, it's time to relive your childhood, providing you were a bit of a nerd in your childhood.

They attract. They repel.
They have a south pole.
They have a north pole.
They're magnets!

3 Chapter

Things to do before you read the science stuff

I try to go out of my way to avoid having you use special equipment in these activities, but there's no avoiding it here. What you need are two bar magnets—the ones that look like Figure 3.1. These might or might not have an N on one end and an S on the other. If you're a classroom teacher, someone in your school will undoubtedly have a couple of bar magnets. If you're not a classroom teacher, try an educational supply store or a hobby store.

Figure 3.1

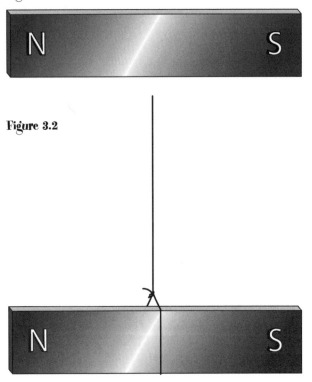

Figure 3.2

While you're out shopping for magnets, get yourself some small, cheap, round or rectangular magnets at a hobby store.

Tie a string around the center of one of the bar magnets so you can hold the magnet by the string while the magnet is free to rotate, as in Figure 3.2. It's tough to get the magnet to balance, so be patient.

Once you have the magnet hanging freely, bring one end of the other bar magnet near and see what happens. Bring the opposite end of the second magnet near the hanging magnet and see what happens.

Take a couple of your cheap magnets and see if they behave the same as your bar magnets (no, you don't have to hang one of them from a string!). They won't behave exactly the same simply because of their shape, but you should still notice both attraction and repulsion.

SCI
LINKS.
THE WORLD'S A CLICK AWAY

Topic: Types of Magnets

Go to: *www.scilinks.org*

Code: SFEM07

You undoubtedly know that magnets don't interact just with other magnets, but with some materials. How else would you display the kids' artwork on the fridge? Even though you already know that magnets stick to refrigerators, take a few moments and find out what other materials are attracted to magnets. And if you think it's as simple as separating things into metals and nonmetals, think again. Try attracting a nickel or a penny with a magnet. While you're messing about with these different materials, see whether it makes a difference which end of a bar magnet you point toward a paper clip or a refrigerator. Does one end attract and the other repel, or do both ends do the same thing?

The science stuff

Let's start with the bar magnets, because their construction makes them the easiest to understand. As long as the bar magnets you used aren't really old and haven't been banged around a lot,[1] they have two ends, known as the **magnetic poles**, which behave differently. We call those the north pole of the magnet and the south pole of the magnet (Figure 3.3). For you really perceptive people out there,

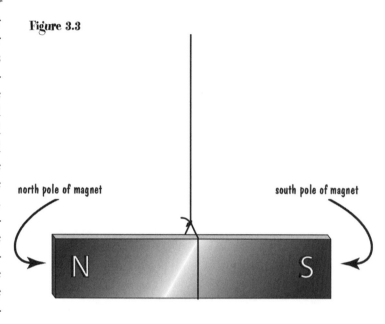

Figure 3.3

north pole of magnet

south pole of magnet

the end of a bar magnet that's labeled "N" is the north pole and the end that's labeled "S" is the south pole.

There's a simple rule for magnets, which is not unlike the rule for how electric charges behave.

Unlike poles of magnets attract each other and like poles of magnets repel each other.

This little rule explains what happened when you brought a second bar magnet near the hanging bar magnet. If the two poles you brought together were unlike poles, say a north pole and a south pole, the two bar magnets lined up. If the two poles you brought together were like poles, the hanging magnet swung around until unlike poles were together, with the magnets lined up again.

---

[1] For reasons I'll discuss in Chapter 4, bar magnets that have been dropped a lot, stored improperly, or otherwise abused might not behave as I'll describe in this section. That's why it might be a good idea to use newer bar magnets or ones that have been "recharged" recently. Why yes, it would have been more considerate of me to mention that when I asked you to get the magnets!

Figure 3.4

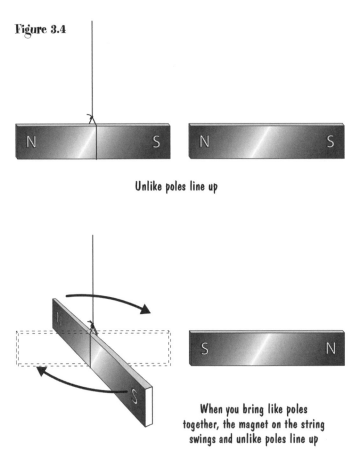

Unlike poles line up

When you bring like poles together, the magnet on the string swings and unlike poles line up

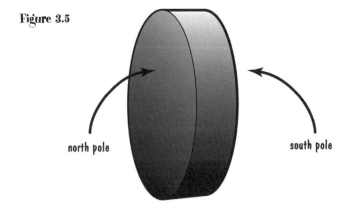

Figure 3.5

north pole   south pole

Figure 3.4 shows what should have happened.

The reason bar magnets are good things to work with is that it's really clear where the north and south poles are on the magnets. Your cheapo magnets might not be so simple, especially if you're dealing with rectangular magnets. Small circular magnets *usually* have opposite poles on each side, as shown in Figure 3.5, but that's not always the case.

Rectangular magnets might have poles located as shown in any of the examples in Figure 3.6, or possibly some other strange configuration. It all depends on how the magnets are manufactured.

Because you did the activities as I asked you to, you also know that magnets attract other materials that are not normally magnetic. Magnets attract all sorts of metals, but not all metals. For

SCI**LINKS**
THE WORLD'S A CLICK AWAY

Topic: Magnetic Materials

Go to: *www.scilinks.org*

Code: SFEM08

example, you would have a difficult time being a pickpocket if your technique involved using strong magnets to attract loose change from people's pockets. If you're after a large paper clip collection, however, this technique will net you a sizeable treasure. Most nonmetals, such as wood, plastic, and glass, just sit there when you bring a magnet near.

Before you move on to more fun things to do with magnets, let me point out that we have *described* the behavior of magnets, but we really haven't come up with anything you might call a scientific model. For static electricity, we started by talking about the existence of different kinds of charges and the fact that they're attracted or repelled. We're at that same stage with magnets, speaking of north and south poles. What we don't yet have is a picture of what the atoms inside magnets might be doing to make them magnetic, or what might be happening inside certain metals for them to be attracted to magnets. We'll get that scientific model in the next chapter.

Figure 3.6

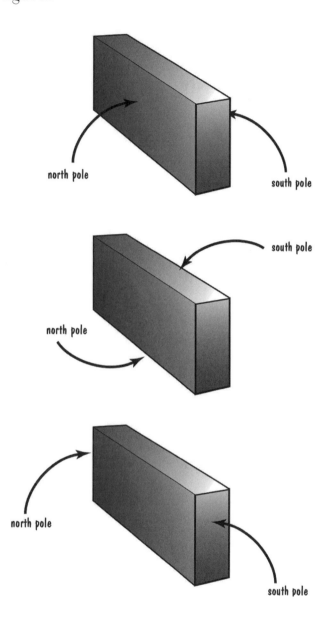

More things to do before you read more science stuff

I'm going to begin this section with a review of something I talked about in Chapter 1. With electric charges, it's possible to completely separate positive and negative charges from one another, as when you cause electrons to jump from one place to another. It's also possible to distort the positive and negative charges in an atom, or have a situation where the positive and negative charges are on opposite ends of the molecule (Figure 3.7). Check back in Chapter 1, and you'll see that we called such an arrangement an *electric dipole*.

Figure 3.7

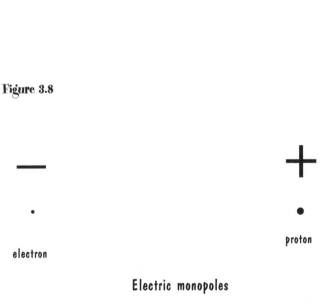

atom

water molecules

Electric dipoles

Figure 3.8

electron

proton

Electric monopoles

Electric charges also can exist as **monopoles**, which are isolated positive or negative charges. A proton all by itself is an isolated positively charged object, so a proton is an **electric monopole**. An electron is an isolated negatively charged object, so an electron is also an electric monopole (Figure 3.8).

Lest you think you misread the heading of this section, I'll now give you something to do. You already know that magnets can exist as dipoles, because the magnets you played with had a north pole and a south pole. Using your cheap magnets, see if you can create a magnetic monopole. In other words, by cut-

SCiLINKS.
THE WORLD'S A CLICK AWAY

Topic: Properties of Magnets

Go to: *www.scilinks.org*

Code: SFEM09

National Science Teachers Association

ting, breaking, chopping, or otherwise mutilating your cheap magnets, can you get a north pole of a magnet all by itself or a south pole of a magnet all by itself? Before you do that, you might ask how you will know if you've isolated a single pole of a magnet. Well, I'll tell you. If you get a piece of magnet that is repulsed by the north pole of a second magnet no matter how you orient the piece, then you have isolated a north magnetic pole. If you get a piece of a magnet that is repulsed by the south pole of a second magnet no matter how you orient the piece, then you have isolated a south magnetic pole. Happy mutilating.

Once you get tired of looking for magnetic monopoles, find yourself a compass—you know, one of those things that's supposed to point North. Any old compass will do, so if you don't have one, head to a party store or a hobby store and buy the cheapest compass you can find. In addition to a compass, you will need a pile of iron filings. These might be a bit more difficult to find at the store, but fortunately, you have an abundant supply outside (provided there isn't a foot of snow on the ground). Just drag your magnets through a bunch of loose dirt, or better, a pile of sand. Your magnets will pick up a whole bunch of iron filings as you do this (Figure 3.9). All you have to do is then use your fingers to scrape the filings off the magnets and into some kind of container.

Figure 3.9

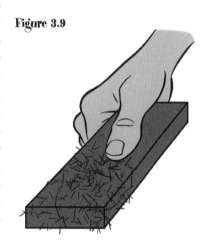

Once you have your iron filing collection, place one of your bar magnets on a flat surface and cover it with a sheet of white paper. Then sprinkle a bunch of iron filings evenly over the top of the sheet of paper, as in Figure 3.10. Lightly tap the paper a bunch of times until the iron filings form a definite pattern. Does this remind you of anything you did in Chapter 2, say something that involved hairs and baby oil? It should!

Figure 3.10

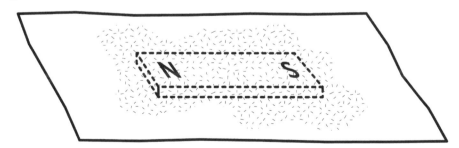

Figure 3.11

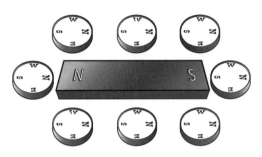

Leave your setup with the magnet and the pattern of iron filings where it is, and place your second bar magnet nearby (but at least a foot away so it doesn't disturb your existing pattern) on a flat surface. Take your compass and move it slowly around the bar magnet, noticing which direction the compass needle points when the compass is in different positions (see Figure 3.11 and note that the compasses there have no needles—no sense giving away the result you'll get).

Compare the direction the needle points at each position with the pattern the iron filings create on the paper covering the other bar magnet.

More science stuff

Let's start with your search for the elusive magnetic monopole. Were you able to isolate either a north magnetic pole or a south magnetic pole by cutting or chopping up a magnet? The answer is no. No matter how you slice and dice a magnet, each resulting piece will still have a north pole and a south pole. Obviously, then, magnets aren't exactly like electric charges. You can give an object an overall negative or positive charge, but you cannot make an object act like only a south magnetic pole or like only a north magnetic pole. There's a good reason for this, but as with so many other things I've teased you with in this chapter, you'll have to wait until Chapter 4 for the explanation!

Figure 3.12

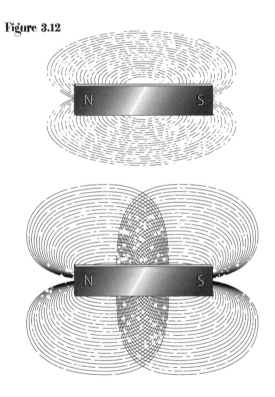

On to the iron filings. If your bar magnets are new or have been stored properly, then you ended up with a pattern like the one on the top in Figure 3.12. If your bar magnets have been around a while and have been abused, you might get a pattern more like the one on the bottom in Figure 3.12.

When you used a charged balloon to cause tiny hairs to line up in a pattern in Chapter 2, we used that to introduce the idea of an *electric field*. Well, now we're going to introduce the idea of a **magnetic field**. We know that magnets exert a force even when they're not touching other objects or other magnets. To explain that, we say that a magnet creates its own magnetic field in the space around it. The magnetic field lines go from the north pole of a magnet to the south pole of a magnet. Figure 3.13 shows the magnetic field lines for a bar magnet and for a circular magnet.

SCI
LINKS.
THE WORLD'S A CLICK AWAY

Topic: Magnetic Fields

Go to: *www.scilinks.org*

Code: SFEM10

To explain how two magnets interact, we simply make up the following rule:

Magnets that are free to move tend to line up with the magnetic field lines created by other magnets. The free magnets line up so their north poles head in the direction of the field lines and the south poles head in the direction opposite the field lines.

Let's see if that explains our observations. First, let's look at what happens when you bring a bar magnet near a second bar magnet that's hanging on a string. We already know what the magnets do, because we saw what they did in the first section of this chapter. The magnet on a string swings around so its north pole is facing the south pole of the second magnet, or so its south pole is facing the north pole of the second magnet. We can explain that result by saying that like poles repel and unlike poles attract, or we can explain it as the magnet on a string lining up with the magnetic field of the second magnet. Figure 3.14 shows how that works.

Figure 3.13

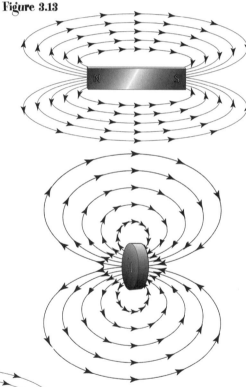

Figure 3.14

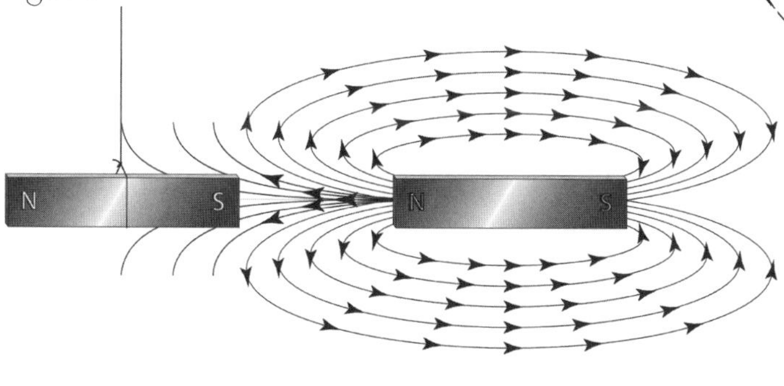

A compass is nothing more than a tiny magnet (the compass needle) that is free to rotate. Therefore, the compass needle will line up with the magnetic field of a bar magnet when you move it to different positions around the bar magnet. Because the direction of the bar magnet's magnetic field is different at different places, the compass needle points in different directions at different positions around the bar magnet. Figure 3.15 shows this.

Figure 3.15

Okay, that explains why a compass needle does what it does when near a bar magnet, but what about the iron filings? The iron filings themselves aren't magnets (try to pick up a paper clip using a bunch of iron filings!), so why do they line up with the magnetic field of a bar magnet? The answer is that when near a magnet, the iron filings act like magnets themselves. This is an example of **induced magnetism**, not unlike instances where you created an *induced* charge in otherwise neutral materials in Chapter 1. How does this induced magnetism happen? Gee, what say we wait until Chapter 4 for that explanation?

Even more things to do before you read even more science stuff

This is an easy "things to do" section. Take your compass and move it away from any bar magnets or telephones or stereo speakers or automobile engines (all those devices have strong magnets in them!). What you should find is that the needle on your compass always points toward geographic North (you know, toward the North Pole of the Earth). Your compass undoubtedly has an N, an S, an E and a W printed on it. Don't expect the needle to point to the N on the compass unless you rotate the compass so that happens. The compass needle will point to geographic North no matter how you orient the letters on the compass.[2]

Topic: Earth's Magnetic Field

Go to: *www.scilinks.org*

Code: SFEM11

Now answer a question. If the needle of a compass is itself a magnet, and magnets line up with the magnetic fields

---

[2] I'm not trying to insult anyone's intelligence with this remark. It's just that I've run into a significant percentage of people who get confused when the compass needle doesn't point to the N printed on the compass!

of other magnets, what's creating the magnetic field that the compass needle is lining up with?

Even more science stuff

You probably were able to answer that last question, but just in case you weren't, the Earth itself is a magnet. Your compass needle is lining up with the magnetic field of the Earth. What causes the Earth to be a magnet? You guessed it—Chapter 4! I can give you a few facts about the Earth's magnetic field, though. First, the Earth's magnetic field is relatively weak compared to the magnetic field of a bar magnet or even cheap refrigerator magnets. Bring any kind of permanent magnet near a compass, and the compass needle will line up with the magnetic field of that magnet rather than the magnetic field of the Earth. Second, the North geographic pole of the Earth is really the *south* magnetic pole of the Earth. That means that the north magnetic pole of a compass needle is the one that points toward geographic North. If the North geographic pole of the Earth were also its north magnetic pole, then the north magnetic pole of a compass needle would point toward the Earth's *South* geographic pole. While you're trying to make sense of that last statement, here's a third fact. If you wanted to find Santa's workshop and did so by following a compass needle North, you wouldn't see any elves. That's because the Earth's north magnetic pole is not exactly in the same place as the Earth's North geographic pole. The two are about 1000 kilometers (600 miles) apart. Armed with all these facts, we can draw a picture of the Earth's magnetic field lines, as in Figure 3.16. Figure 3.16 shows magnetic north as being a bit farther than 1000 kilometers from geographic North, and it shows magnetic north as being maybe in the northern part of the Asian continent. As of this writing, magnetic north is somewhere in Canada and moving toward

Figure 3.16

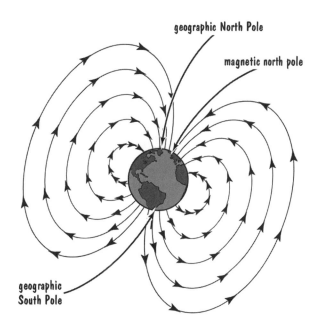

Earth's magnetic field lines

Greenland. No big deal about not having the location exact, because the location *does* change. Magnetic north generally moves about 15 kilometers per year.

Although the Earth's magnetic field is rather weak, it still leads to interesting occurrences, as I'll explain in the Applications section and in later chapters.

Chapter Summary

- Permanent magnets have a north pole and a south pole. Like poles of magnets repel and unlike poles of magnets attract.

- Permanent magnets can *induce* magnetism in some metals, causing the metals to be attracted to the magnet.

- Magnets exist as dipoles, and no one has been able to isolate a magnetic monopole.

- We can describe magnetic interactions using the concept of a *magnetic field*. Magnets produce magnetic fields, and other magnets tend to line up with these magnetic field lines.

- A compass contains a small, magnetized needle that is free to rotate. In the absence of stronger magnetic fields, a compass needle will line up with the Earth's magnetic field.

- The Earth creates its own magnetic field. The magnetic poles of the Earth are not exactly in the same locations as the Earth's geographic North and South poles.

Applications

1. Ever wonder how homing pigeons manage to find their way home even though they, like human males, never stop to ask directions? There's strong evidence to suggest that homing pigeons rely on the magnetic field of the Earth to know where they are and how to get home. How in the world do we know this? Well, the Sun sometimes emits *solar flares* that really mess up the magnetic field lines of the Earth. During the emission of these solar flares, homing pigeons do a good job of getting lost. Certain atoms and molecules in the brain of a homing pigeon act like tiny magnets and are able to line up with the Earth's magnetic field. Don't ask me how a bird with a tiny brain can retrace the various changes in alignment with the Earth's magnetic field as the bird goes from one location to another, but apparently that's what they do! It's also thought that whales use a similar technique, among other methods, to navigate during their yearly migrations.

2. There is a definite tendency for human males to rely on geographic methods for getting from one place to another and for human females to rely more on landmarks when they navigate. In other words, males tend to have a general sense of direction while females tend to use street names and other identifiable features to figure out how to get someplace. That's not sexist; it's just the way things are. Well, it turns out that there's a biological basis for this. Humans have a small number of very tiny iron filings in the bone and cartilage of the nose, right in the center near the eyes. There is evidence that people can use these filings to orient themselves toward magnetic north even when blindfolded. Males tend to have more of these filings, which explains why males tend to be better at geographic orientation. And yes, that really is amazing!

3. Because magnets can repel each other, you might think that you could use magnets for levitation. You'd be right. There are trains in Europe and the Orient that use electromagnets (I'll explain what those are in the next chapter) to levitate the trains above the track. Because the trains don't actually touch the track, there isn't any friction between the train and the track. That makes for a smooth and fast ride.

Connecting Electricity and Magnetism

I'm sure it seemed every time you turned a corner in Chapter 3, I was telling you I'd explain something when we got to Chapter 4. Well, here we are. I also told you in the beginning of the book that there is an intimate connection between electricity and magnetism, so we might as well address that one here, too. We'll end up with a scientific model for what magnetism is, what makes something a permanent magnet, and why magnets attract some metals and not others. Wheee!

Things to do before you read the science stuff

You need to get more specialized equipment for this chapter, but not so much that it will break your budget. First, find a toy train or race-car set transformer. This transformer should have two separate wires that connect the transformer to the train or race-car track (don't worry, you don't need any track!). The wires that come with the transformer should be "stripped" on both ends, so you have a small length of bare wire on each end that is not covered with plastic insulation. You also need a bunch of metal paper clips; a nail (size 16d works well, but the size isn't critical); a **60** centimeter (approximately) length of insulated wire that's about 22 or 24 gauge (go to the hardware store and ask for "bell wire"); a 2-meter length of the same kind of wire; and a 1.5 volt battery of any size (AA, C, D, whatever).

Topic: Electrical Safety

Go to: *www.scilinks.org*

Code: SFEM12

Caution: In what follows you are using a toy transformer because the low voltage that is produced by the transformer won't hurt you even if you touch bare wires. *Never* use connections directly from the wall socket of your home to do this or any other activity in this book. Aside from the fact that the electric current direct from the wall socket won't work in this activity, it would be just plain silly to electrocute yourself in the name of learning science.

Figure 4.1

Get the toy transformer but don't plug it in. Twist the two wires together at the end so they make a loop, as in Figure 4.1.[1] Here I'm talking about the bare bits of wire at the end, and not the insulated sections of the wires. Then bend the wires until they're really close, but not touching, at some point. This point should be off the ground or table on which you have the transformer. This is also shown in Figure 4.1. The wires should be stiff enough that they will stay in this position without you holding them there.

Now make sure the transformer is in the off position and plug in the transformer. While watching closely the spot where the wires are the closest, turn the transformer on and off. Look for any motion of the wires as you do this. They won't move a lot, so you have to be a good little observer! Be sure not to leave the transformer in the on position for too long, or the wires will get pretty hot.

---

[1] If you're using a train transformer, there are two sets of connections on the transformer. One set goes to the train track and the other set is used for accessories. Make sure you use the set of connections intended for the train track.

Take your plain nail and put it next to a metal paper clip. Does the nail act like a magnet and pick up the paper clip? Of course not! Now get your 60 centimeter long piece of wire and strip about 2 centimeters of insulation off each end. You can most likely do this with your fingernails, but a knife or scissors might help. Next wrap this wire neatly around the nail, keeping the wraps as close together as possible and leaving maybe 8 centimeters of wire free at each end. Check out Figure 4.2 if you have no idea what I'm talking about.

Now touch the free ends of the wire (the metal parts that you stripped of insulation) to opposite sides of the battery as shown in Figure 4.3. Point either end of the nail at a paper clip and see if you can now pick up a paper clip. You should be able to.

Figure 4.2

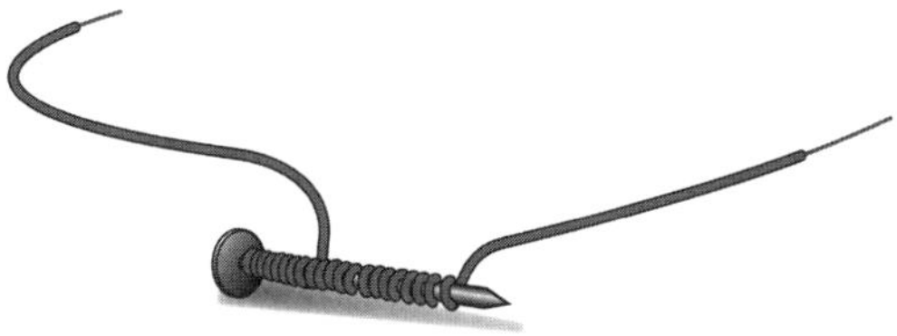

To see that you've turned your nail into a magnet by connecting the wire to a battery, bring the north or south pole of one of your bar magnets near one end of the nail and then near the other. As long as you have the battery connected to the wire, your nail should attract the north pole of the magnet on one end and repel it on the other. See Figure 4.4.

Figure 4.3

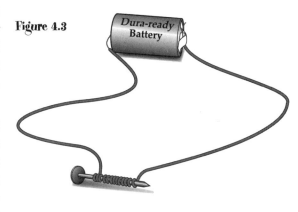

For further evidence that your nail wrapped in wire is a magnet, tape the ends of the wire to the battery so you don't have to hold them in place. Then place this contraption on a flat surface and place a blank sheet of paper over the nail. Sprinkle iron filings on top of the paper and tap, as you did

Figure 4.4

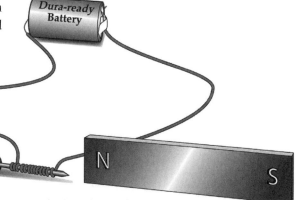

You get attraction at one end and repulsion at the other

with the bar magnet in Chapter 3. Does the resulting pattern look familiar?

Remove the wire from your nail and see if the nail alone can pick up a paper clip, or at least attract it. You should notice that your nail is still acting at least a little bit like a magnet. Go outside and throw the nail down on a hard surface, such as a sidewalk, several times. Don't be shy—bang that nail hard! Then go back inside and see whether or not the nail still acts like a magnet and attracts paper clips. Hmmmm.

Now take one of your cheap magnets and get a pile of metal paper clips. See how many paper clips you can pick up with the magnet. Then take your cheap magnet outside and treat it as poorly as you did the nail. Throw it down on a hard surface three, four, or twenty times. Go back inside and see how many paper clips the magnet will pick up now. Do you suppose the magnet is just tired from being beat up?

The science stuff

Hopefully you noticed that when you ran an electric current through the wires leading from your transformer, they moved ever so slightly toward each other and then returned to their original positions when you turned the transformer off. If they didn't do that, go back and try the activity again with the wires even closer together, making sure the wires are free to move and not caught on something. What we learn from this is that *electric currents can exert forces on each other.*[2] In this case, the currents were moving in opposite directions (see Figure 4.5) and they

Figure 4.5

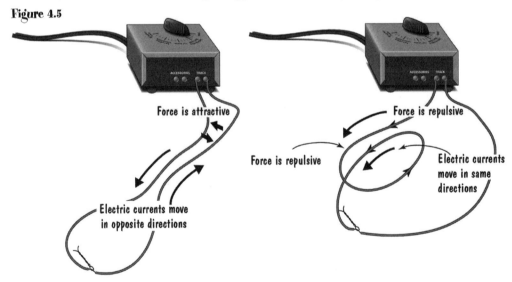

Force is attractive

Force is repulsive

Force is repulsive

Electric currents move in same directions

Electric currents move in opposite directions

---

[2] In case you've forgotten, electric currents are defined as moving charges. The charged objects that usually do the moving are electrons moving through metals.

attracted each other. If the currents were moving in the same direction, they would repel each other. You can check this out by rearranging the wires as shown in the second drawing in Figure 4.5.

SCI LINKS.
THE WORLD'S A CLICK AWAY

Topic: Electric Current

Go to: *www.scilinks.org*

Code: SFEM13

Okay, so electric currents exert forces on each other. What if we took two coils of wire, as shown in Figure 4.6, and ran electric currents through them. Would the two current-carrying wires exert forces on each other? Sure they would.

Figure 4.6

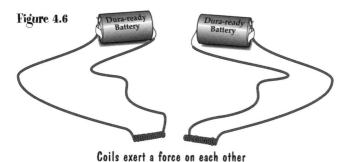

Coils exert a force on each other

To actually see this happen, you would have to run rather large currents through the wires, and since I don't want you to electrocute yourself, I won't have you do that. Instead, you could insert a nail inside each coil of wire. That would make the effect of the currents much stronger (I'll explain why in a bit), and the two coils of wire plus nails would exert forces on each other. You can test that, of course, by constructing two nails wrapped with wire and noticing that they can repel and attract each other when each is connected to a battery.

What we have so far is that wires carrying electric currents exert forces on each other, and that also applies to coils of wire carrying electric currents. We also know, however, that coils of wire carrying electric current act like magnets. You saw that when your nail wrapped with wire picked up metal paper clips and when it repelled and attracted a permanent magnet. Again, you'll have to trust me, but with larger electric currents, your coil of wire would act like a magnet even without the nail inside.

Ready for the big connection between electricity and magnetism? The presently accepted model of magnetism is that it's nothing but electric currents. In other words, electricity and magnetism are the same thing! If it looks like a magnet, acts like a magnet, and quacks like a magnet, then it must be a magnet, even if it's an electric current.

One of the main goals of physicists is to try to "unify" concepts, meaning to look for similarities in seemingly separate observations and try to explain those separate observations with one scientific model. The early Greeks knew about electrostatic effects and they were aware of magnets. It wasn't until 1820, however, that a physicist named Hans Christian Oersted (no, he didn't write physics fairly tales) established a connection between the two phenomena. Later in the 1800s, one of the great breakthroughs in physics came when someone named James Clerk Maxwell provided a set of mathematical relationships that showed the almost exact symmetry between electricity and magnetism. In other words, he showed that electricity and magnetism were two aspects of the same thing. Physicists now speak of one concept, which is **electromagnetism**, rather than separate concepts of electricity and magnetism. Today, physicists are searching for one theory or model that will tie together all the kinds of forces that we know of (gravitational, electromagnetic, strong, and weak forces[3]). So far, we have a pretty good theory that unifies the electromagnetic force and the weak force, but the strong force and the gravitational force don't yet fit into that picture. Yes, there are many unanswered questions, meaning that you too can still make a major contribution to the body of scientific knowledge!

Figure 4.7

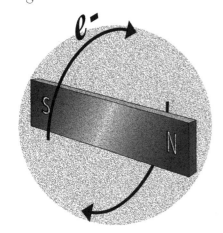

To understand the presently accepted model of magnetism, we have to get down to the level of atoms—those things that are composed of protons, neutrons, and electrons. The model of atoms you probably learned in school was one with a positive nucleus (the protons and neutrons) surrounded by electrons orbiting the nucleus. I already told you in Chapter 1 that this model is incorrect, but let's go with it for now because it *does* predict observations correctly. If electrons are orbiting in circles around the nucleus, then you have an electric current.[4] An electric current in the shape of a circle acts just like a magnet, as you saw with the coils of wire surrounding the nail. So, each atom individually acts like a magnet, with a south pole and a north pole (Figure 4.7).

---

[3] I haven't mentioned the "weak" force before. It's a force that is associated with the release of electrons from nuclei and the interactions between electrons and particles that are similar to electrons. And yes, I haven't begun to tell the story of the current theories about what goes on at the subatomic level. There are many, many particles involved in our present view of that tiny world. And you thought we just had protons, neutrons, and electrons. Wrong!

---

[4] Remember, we define an electric current as charged objects that are moving. Electrons moving in circles fit that definition.

Because all materials are composed of atoms, all materials act like magnets, right? Well, no. For a material to be magnetic, most of the electrons in the atoms of the material would have to be orbiting in the same sense (clockwise or counterclockwise) and oriented toward the same direction, as shown in Figure 4.8.

In most materials, the tiny little magnet atoms are randomly oriented, so that the effects of the separate little magnets cancel. If you have a large collection of randomly oriented magnets, they would not be magnetic as a whole even though each individual magnet would be magnetic. See Figure 4.9.

Before we go any further, I really need to be more accurate in describing what's going on with atoms in materials. As I said, electrons in atoms are *not* orbiting along circular paths. We know that's the case because if electrons orbited in circles, they would radiate away energy rapidly and collapse into the nucleus of the atom. If that happened, the world as we know it would collapse upon itself in a very short time. Because that doesn't seem to be happening, we make the bold assumption that electrons don't move in circular orbits. What electrons *are* doing, we don't really know. The best model we have at present is that electrons can be found in something called a **probability cloud** that surrounds the nucleus. I'm sure that's

Figure 4.8

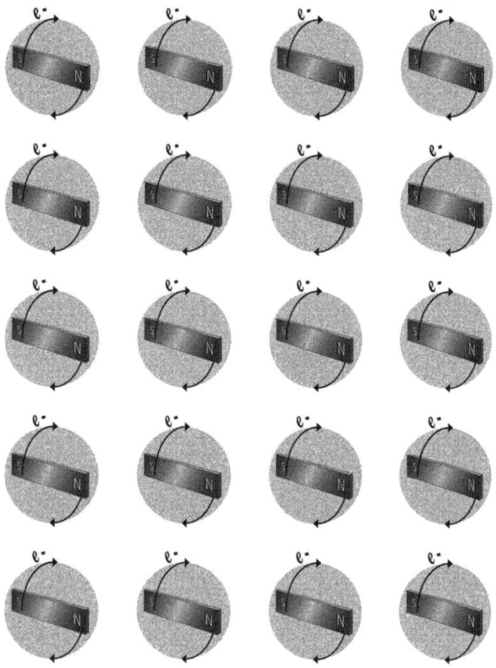

Figure 4.9

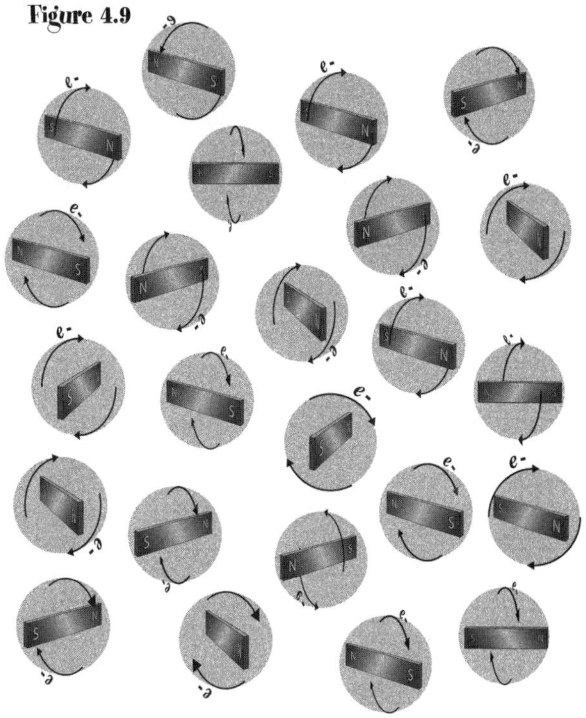

more than you want to know (or is it less than you want to know?) about what electrons do, but I do need to add an important point: for the purposes of describing the magnetic properties of atoms, electrons *act* as if they are moving along circular paths, creating a circular electric current! To get around that difficulty, we describe electrons as having something called a **magnetic moment**, which is not to be confused with a senior moment.[5] Basically, a magnetic moment can be thought of as a tiny magnet. Protons and neutrons also have magnetic moments, so what determines whether or not a particular atom acts like a magnet depends on how the magnetic moments of its electrons and protons and neutrons add together.[6] If the magnetic moments don't cancel one another and are unbalanced, then an atom can be thought of as behaving like a magnet. If the magnetic moments do cancel, then the atom as a whole will not act like a magnet.

We're still not quite at an accurate picture of the presently accepted model of magnetism, but we're at a point where you can get an understanding of what makes something a permanent magnet, why magnets can attract certain materials, why you can cause unmagnetized things like nails to become magnetized, and why you can ruin a magnet by throwing it on a sidewalk. Don't worry, though. For you purists out there, I'll present a more accurate model of magnetism in the Applications section of this chapter. Why the Applications section? Because we have too many consecutive pages of explanation going as it is, and I don't want you to get the feeling you're sitting in a big lecture hall counting the concepts as they fly over your head!

Okay, let's look at the atoms in a substance as a big array of magnetic moments that might or might not be free to rotate or at least flip between two directions. If those magnetic moments are randomly oriented or are half in one direction and half in the opposite direction, the material will not be magnetic because the magnetic moments cancel one another. See Figure 4.10.

---

[5] In case this doesn't make sense, a senior moment is when your mind goes blank for a while. That's pretty much what happened when I tried to think of a "joke" moment to use here. I asked my wife what example to use and she drew a blank. For some reason, "senior moment" came to mind.

[6] The complete picture of magnetic moments in atoms can get a bit complicated, as there are orbital magnetic moments and what are called "spin" magnetic moments. No need for us to go into that here, so we'll just pretend the only things making atoms act like magnets are the magnetic moments of the electrons. And by the way, I add comments like this one so you're aware of when the scientific models I'm presenting to you are somewhat simplified from the actual models. I'm not trying to overwhelm you and I certainly don't expect you to understand the complete scientific model based on a few sentences.

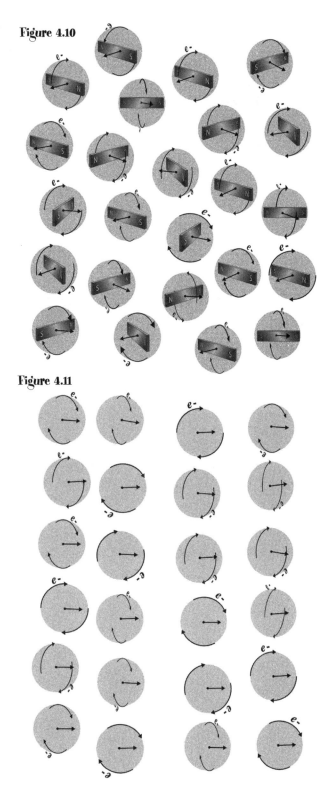

If the magnetic moments of all the atoms more or less line up, though, they reinforce one another and make the material as a whole magnetic, as in Figure 4.11.

So, we have a picture of what's going on in a permanent magnet. The magnetic moments of the individual atoms in a permanent magnet are lined up and they tend to stay that way. The more lined up the magnetic moments of the atoms, the stronger the magnet. The strength of the magnet is also affected by the size of the magnetic moments of the individual atoms. Large magnetic moments of the atoms lead to stronger magnets. This also explains why some materials make lousy magnets. Most nonmetals have atoms with tiny or even almost nonexistent magnetic moments.[7] You can line up these magnetic moments all you want, and you still won't get much of a magnet.

Let's look at what happens when you bring a permanent magnet near a paper clip. Before you bring the permanent magnet near, the magnetic moments of the atoms in the paper clip are pointed in random directions. These magnetic moments are free to rotate, however. When you bring the permanent magnet near, the magnetic

Figure 4.10

Figure 4.11

[7] In case you love vocabulary words, substances that contain atoms with almost nonexistent magnetic moments are called **diamagnetic**.

moments of the atoms in the paper clip tend to *line up* with the magnetic field of the permanent magnet (we learned about this in Chapter 3). Once those magnetic moments line up, the paper clip itself acts like a magnet and is attracted to the permanent magnet. You could say that the paper clip has **induced magnetism**. Take the permanent magnet away, and the magnetic moments of the atoms in the paper clip return to a random orientation, leaving the paper clip nonmagnetic.[8] Figure 4.12 illustrates this process.

Figure 4.12

Magnetic moments randomly oriented

Magnetic moments line up in the presence of a magnet

Magnetic moments random again after magnet is removed

Why won't a permanent magnet attract all materials, including some metals? Because some materials have either small magnetic moments to begin with, or they have atoms whose magnetic moments do not easily change direction. If you can't line up a bunch of large magnetic moments, you can't make an object act like a magnet.

All right, let's look at that nail that you made into a magnet. Before you sent an electric current through the wire wrapped around the nail, the nail didn't act like a magnet. After you connected the wire to the battery, though, the nail could pick up a paper clip or two. To understand this, remember that we are assuming that a circular electric current *is* a magnet. Therefore, the coil of wire wrapped around the nail creates a magnetic field that looks like Figure 4.13.

Figure 4.13

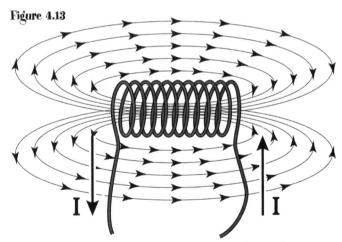

Magnetic field lines from a coil carrying an electrical current

---

[8] More vocabulary if you're interested. Substances that contain atoms with reasonably large magnetic moments that will line up with an external magnetic field and then return to a random orientation when the field is removed are called **paramagnetic**.

The magnetic moments of the atoms in the nail see this magnetic field from the coil of wire and line up with it. By lining up with the magnetic field of the wire, these magnetic moments make for a *stronger* magnetic field (Figure 4.14). Thus, a coil of current-carrying wire with a nail inside is a stronger magnet than just the coil of wire alone.

Figure 4.14

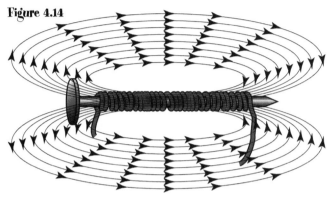

Magnetic field lines from a nail wrapped in a coil carrying an electrical current

When you removed the coil of wire from the nail, the nail was still slightly magnetic. That's because once lined up, the magnetic moments of the atoms in the nail don't head back to a random orientation as easily as do the magnetic moments of the atoms in a paper clip. The magnetic moments in the nail are still somewhat lined up, and still make the nail act like a magnet.[9] How to get those magnetic moments randomly oriented again? Simple. Bang the heck out of the nail! That shakes things up inside the nail and gets those magnetic moments into a more random orientation. Of course, that method for demagnetizing something works not just for nails, but also for "permanent" magnets, as you saw if you took my advice and threw a cheap magnet onto a hard surface several times. Whack a magnet around enough and it won't be a magnet for long. You can also ruin a magnet by heating it, as high temperatures make atoms move around more, messing up the orientation of their magnetic moments. Of course, it is possible to realign the magnetic moments in a magnet by placing it in a strong magnetic field. So, all is not lost if you have bar magnets that have been banged around and messed up. You can buy a device that will "recharge" bar magnets and get them back to their original condition.

Maybe you've forgotten, but in Chapter 3 I asked you to cut up a cheap magnet in an attempt to isolate either a north or a south magnetic pole. You couldn't do that, as each tiny piece you created had both a south and north magnetic pole. Given our model of what causes magnetism, that now makes sense. If each individual atom is itself a magnet, then no matter how small a piece you get from a

Topic: Electromagnetism

Go to: *www.scilinks.org*

Code: SFEM14

---

[9] Even more vocabulary. Substances containing atoms whose magnetic moments tend to stay lined up when an external magnetic field is removed are called **ferromagnetic**.

magnet, that tiny piece is still a collection of tinier magnets (the atoms themselves). In fact, since each atom is a magnet, you would have to begin cutting atoms apart in your search for a magnetic monopole. Physicists never have been able to isolate a magnetic monopole, even by breaking atoms apart, so if you do manage to find one, call your local physicist—your Nobel Prize is waiting.

Because you are undoubtedly on "explanation overload," maybe it's time for you to play around a bit more. Sounds good to me, but first a recap is in order. Electric currents exert forces on other electric currents. Electric currents that move in a circle act like permanent magnets. We use those facts to create a model of magnetism that includes individual atoms acting like they are circular electric currents, which we call magnetic moments. The orientation or lack of orientation of the magnetic moments of atoms explains why permanent magnets do what they do, why magnets attract some materials and not others, and why you can ruin magnets by hitting them or heating them.

More things to do before you read more science stuff

I'm sure you're ready for a break, so turn on a television that has a picture tube. Projection TVs are no good for what you're about to do. Bring one of your cheap magnets near the screen of the TV,[10] and move it around a bit. You should notice that the picture is distorted near the magnet. That's all for this section, unless you want to do a bit of research and find out how a television produces a picture. If you aren't in the mood for research, head for the next section and I'll tell you how televisions work. And yes, that's all you have to do in this "things to do" section. Exhausting, wasn't it?

More science stuff

Inside the picture tube of a television set is an electron gun that fires electrons at the screen. That's right, I said an electron gun. The way it works is that a heated coil at the back of the picture tube produces electrons (heated coils do that). The front of the picture tube acts like it's positively charged (I'll explain how when I address *voltage* in Chapter 5). When electrons leave the back of the tube, they zoom toward the positively charged screen, hitting things called *phosphors* that produce light. Don't worry if that brief explanation isn't crystal clear. What's important is that you have electrons traveling from the back of the picture tube to the front.

---

[10] Do *not* use a strong magnet for this! You could ruin your television set, and that might be a bad thing, depending on your view of what comes over the airways.

When you place a magnet near the screen, you deflect the electrons from their intended path, leading to a distorted picture at that point. That leads us to the following conclusion:

A magnetic field exerts a force on moving charges.

That might seem like a new idea, but it's not. We saw at the beginning of this chapter that electric currents exert forces on other electric currents. Your magnet that creates the magnetic field is nothing but a collection of many tiny electric currents (or at least magnetic moments that act like electric currents). An electric current consists of moving charges, so a bunch of moving charged electrons constitutes an electric current. Therefore, saying that the moving electrons inside the TV experience a force when in the presence of the magnetic field of the magnet is really just the same as saying that one electric current exerts a force on another.

Even though I haven't really given you a new explanation, this new way of looking at things gives us a slightly new perspective on electric currents and magnetic fields. Let's go way back to when you observed two current-carrying wires move together. In one wire, you have moving charged objects (electrons). If those electrons experience a force (which they do), then there must be a magnetic field present, right? But what causes the magnetic field? The electric current in the other wire, of course. You see, we can generalize our previous statement that electric currents moving in circles act like magnets to the following: *electric currents generate magnetic fields*. There's a neat little trick to figuring out what the magnetic field produced by an electric current looks like, but I'll spare you that since it won't do a lot for your understanding right now.[11] Instead, I'll just show you what the magnetic field created by a current-carrying wire looks like. That would be in Figure 4.15.

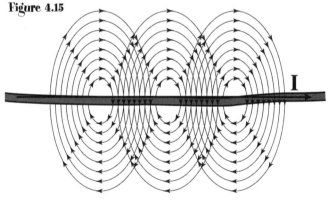

Figure 4.15

Magnetic field lines from a wire carrying an electric current

---

[11] If you're interested in the trick, pick up any high school or college physics text and look up the "right-hand rule." There's one right-hand rule for figuring out the magnetic field produced by an electric current and another right-hand rule for figuring out the direction of the force exerted on a moving charge by a magnetic field.

So, we have two wires next to each other. Each wire carries an electric current. The current in one wire creates a magnetic field around that wire. The electric current in the other wire sees that magnetic field and feels a force, because magnetic fields exert forces on moving charges. Of course, this also works in reverse. The electric current in each wire reacts to the magnetic field produced by the other wire. There's a force exerted on each of the wires (Figure 4.16).[12]

Figure 4.16

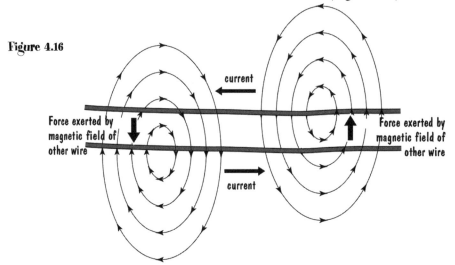

Magnetic field lines for carrying currents in opposite directions

Before moving on, take a moment to put everything in perspective. We start with a basic observation, which is that two wires connected to a battery or a toy transformer push or pull each other. To explain this observation, we invent the idea of tiny electrons moving through the wires (we call this an electric current). We further invent the idea that moving electrons generate a magnetic field that we can't see. Other moving electrons (the current in the other wire) see this magnetic field and feel a resulting force. You can believe that moving electrons and magnetic fields exist, or you can believe that they don't exist. What you believe isn't important. What's important is that the scientific model consisting of moving electrons and magnetic fields successfully explains what we observe.

I'd like to point out the parallels between our models of electricity and magnetism. The mere presence of an electrically charged object creates an electric field (see Chapter 2). Other electric charges see this electric field and feel a force that is dictated by the strength of the electric field. The mere presence of moving charges creates a magnetic field. Other moving charges see this mag-

---

[12] For you Newton groupies out there, this is an example of Newton's third law, which states that when one object exerts a force on a second object, the second object exerts an equal and opposite force back on the first object. For a whole chapter devoted to Newton's third law, take a look at the *Stop Faking It!* book on Force and Motion.

netic field and feel a force that is dictated by the strength of the magnetic field. That's our model and we're sticking to it!

If formulas with lots of letters get you upset, then skip the rest of this section. What I'm about to tell you isn't essential for understanding the rest of the book. On the other hand, I really think I should give you a feel for how high school and college textbooks, and *some* elementary and middle school textbooks, describe the interactions between moving charges and magnetic fields. The relationships I'm about to provide are valuable, because they describe not only *that* moving charges exert forces on one another, but how big and in what direction those forces are. In what follows, I'll be using letters to represent various quantities. q will stand for the amount of charge, measured in coulombs, possessed by an object such as an electron. This is the same q that is part of Coulomb's law, introduced back in Chapter 2. B represents the magnitude of a magnetic field.[13] The larger B is, the stronger the magnetic field. When B is written in boldface, as in **B**, that simply means that the *direction* of the magnetic field is included. A quantity that has both magnitude and direction is known as a **vector**. The other symbol I'll use a lot is I, which stands for the magnitude of an electric current. Electric current is measured in amperes, or just "amps" for short. The magnitude of an electric current is determined by how much charge passes a given point per second, so one ampere is equal to one coulomb per second—charge per unit time. Electric current also has a direction associated with it, so you'll see the symbol **I** as well as the symbol I. And no, I have absolutely no idea why we use I to represent electric current!

Many physical quantities have both a magnitude and a direction. For example, the direction you move is often as important as how fast you move, so we use a vector to represent the combination of those two things and call it *velocity*. Forces also have a direction associated with their magnitude (which direction someone pushes you when you're standing on the edge of a cliff has definite consequences), so force is a vector. From here on out, whenever you see a single letter representing a quantity, and that letter is in boldface, then you're dealing with a vector. Whenever the letter is *not* in boldface, then you're dealing with just a magnitude and not an associated direction.[14]

---

[13] The proper term for B is the *magnetic induction* or the *magnetic flux density*, but most books, including this one, refer to it as the strength of the magnetic field. There is another quantity, represented by the symbol H, which is correctly identified as the magnetic field strength, and you might see H used instead of B in some books. Although they are not exactly the same thing, you won't go too far wrong if you think of B and H as being the same thing.

---

[14] Just to confuse things a bit more, I'll also let you know that a letter with an arrow over it, as in $\vec{a}$, also represents a vector. We won't be using those in this book because it's too much of a pain to create letters with arrows over them!

For starters, we'll deal with the notion that moving charges (electric currents) create magnetic fields. There's a powerful relationship that helps you figure out **B** given any old electric current or collection of electric currents. Because that relationship involves all sorts of symbols used in calculus, and I don't want to have you put this book down now and never pick it up again, I'll just provide a couple of results that one can derive from that powerful relationship. The first is the magnitude of the magnetic field that surrounds a long, straight, current-carrying wire. Here it is:

$$B = \frac{\mu_0 I}{2\pi r}$$

where μ_0 represents a number that is a constant for our purposes, 2 is a number (duh!), and π is that familiar number 3.14159... . Because those three factors all combine to make just one number, we can ignore them in deciding whether or not this relationship makes sense. What we're left with is I, which represents the magnitude of the current flowing through the wire, and r, which represents the distance you are away from the wire. Figure 4.17 should make it all clear.

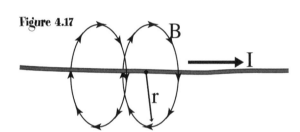

Figure 4.17

Because I is in the numerator of our expression, that means the larger I is, the stronger the magnetic field created. That makes sense, because you would expect bigger currents to exert larger forces on other currents. Because r is in the denominator of our expression, the larger r is, the weaker the magnetic field. All that means is that the farther you are away from the wire, the weaker the strength of the magnetic field. That also should make sense because the force between two wires is stronger when they are close than it is when they're farther away.

Now that you're geared up for understanding formulas that describe the strength of magnetic fields, here's another one. The strength of a magnetic field at the *center* of a loop of current-carrying wire is given by

$$B = \frac{\mu_0 I}{2r}$$

where μ_0 is the same number as in the previous relationship and 2 is, well, the number 2 again. I represents the magnitude of the electric current flowing through the loop of wire and r represents the radius of the loop (which, if you think about it, is again the distance you are from the wire). This setup is shown in Figure 4.18.

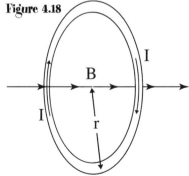

Figure 4.18

I hope again that this formula makes sense to you. If the current I is stronger, it creates a stronger magnetic field at the center of the wire. The larger the loop, the bigger r is. Because r is in the denominator of our expression, a larger r means a weaker magnetic field at the center, which makes sense because the wire carrying the electric current is farther away.

Before moving on to more formulas, I want to repeat that the only reason I'm presenting these here is that you might run into them in various textbooks or other places where they explain electromagnetic interactions. When that happens, I want you to have an idea of the theory behind the formulas, and some idea of what all the symbols represent. That said, let's move on to figuring out the force that a moving charge feels when it encounters a magnetic field. That relationship is given by

$$\boldsymbol{F} = q v \times \boldsymbol{B}$$

In this relationship, everything in boldface is a vector. $\boldsymbol{F}$ is the force felt by the charged object. q is the magnitude of the charge on the object, v is the velocity (speed plus direction) of the object, and $\boldsymbol{B}$ is the strength of the magnetic field the object sees. You might notice that the symbol for multiplication, ×, is also in boldface. That's because this is a special multiplication known as a *cross product*. The cross product contains within it information about the direction and the size of the force the charged object feels. As I mentioned earlier, the direction of the force a magnetic field exerts on a charge is a little screwy, and there's a trick to figuring out that direction. The cross product takes that into account, and that's all you really need to know.[15] The important thing is that the rest of the relationship makes sense. The larger q is, the larger the force. All that says is that objects with a greater charge feel a larger force than objects with a lesser charge. Also, the larger $\boldsymbol{B}$ is, the larger the force. Stronger magnetic fields lead to stronger forces. See Figure 4.19.

Figure 4.19

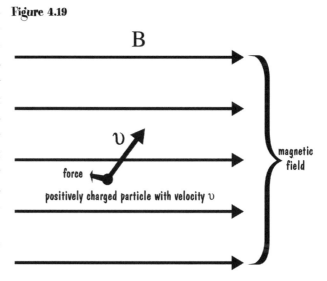

B

magnetic field

force

positively charged particle with velocity υ

---

[15] For those of you who aren't satisfied with this description, I'll give you a bit more information. The cross product between v and $\boldsymbol{B}$ means that the force is a maximum when v and $\boldsymbol{B}$ are perpendicular, and the force is zero when v and $\boldsymbol{B}$ are parallel. In other orientations, the force is between the maximum value and zero.

Just one more formula to give you, which is the formula that tells how big a force a current-carrying wire feels when it sees a magnetic field. This isn't really a new formula, because current-carrying wires consist of moving charges, and we already have a formula for the force exerted on a moving charge by a magnetic field. Anyway, here it is:

$$\boldsymbol{F} = I\boldsymbol{l} \times \boldsymbol{B}$$

Again we have that special multiplication called a cross product, and the magnetic field strength $\boldsymbol{B}$ is the same as in the previous relationship.[16] I represents the magnitude of the electric current in the wire, and $\boldsymbol{l}$ is the length of the wire.[17]

Okay, done with formulas for a while. We won't be using these formulas, but now at least maybe you won't faint when you run into them elsewhere.

Even more things to do before you read even more science stuff[18]

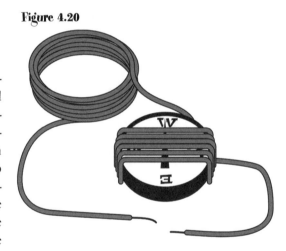

Figure 4.20

For this section, you need your compass, the 2-meter length of insulated wire I had you get earlier, a bar magnet, a battery, and an empty toilet paper tube. Strip a small bit of insulation off each end of the wire and then wrap the wire lots of times around the compass. Leave about 10 centimeters of wire at one end and at least a meter of wire on the other end. After making sure you can still see the compass needle underneath the wrapped wire, tape the wire in place. Take a look at Figure 4.20.

---

[16] As with the previous cross product, the force is a maximum when $\boldsymbol{l}$ and $\boldsymbol{B}$ are perpendicular and zero when they're parallel.

---

[17] In this relationship, $\boldsymbol{l}$ is in boldface. Don't ask me why, but it's conventional to use $\boldsymbol{l}$ as a vector whose direction is in the direction of the electric current. Seems to me it would make more sense to have the electric current I be the vector, but what do I know?

---

[18] This activity is identical to one contained in the *Stop Faking It!* book on Energy. If you did the activity when going through that book, fine. If not, it's about time you got busy!

Make sure the compass isn't near your magnet and place it on a flat surface. The compass needle should point North, and you should know why! Rotate the compass until the needle is lined up with the coil of wire, as in Figure 4.21.

With the compass in that position, touch the free, stripped ends of the wire to the battery terminals, as shown in Figure 4.22. The compass needle should deflect, and you really should be able to explain why, given all you know about electric currents, magnetic fields, and compasses. Even if you have trouble explaining this effect (review this chapter if you do!), at least maybe you can admit that if the compass needle deflects, that means an electric current is flowing through the wire.

Set the battery aside and cut the toilet paper tube in half, as in Figure 4.23. Wrap the long free end of the wire around one of the half tubes as shown in Figure 4.24. Leave enough wire so you can twist together the stripped ends and be able to place the tube at least 30 centimeters away from the compass.

Move the magnet in and out of the center of the tube. As you do this, watch the compass needle (see Figure 4.25).

The compass needle should move just a tiny bit and only while the magnet is moving. Don't expect the needle to move as much as it did when you had the battery connected to the wires.

Figure 4.21

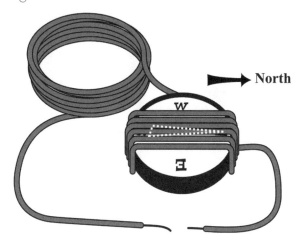

North

Figure 4.22

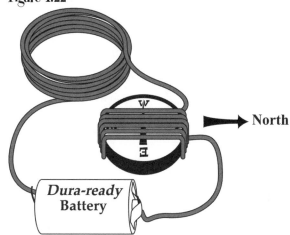

North

Dura-ready Battery

Figure 4.23

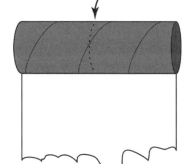

Cut roll in half

Figure 4.24

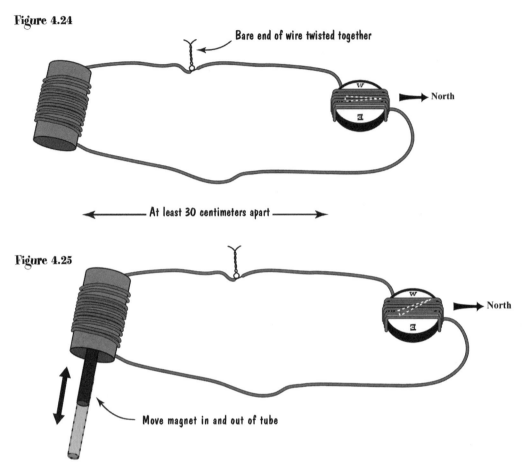

Bare end of wire twisted together

North

← At least 30 centimeters apart →

Figure 4.25

North

Move magnet in and out of tube

Even more science stuff[19]

A magnet has an associated magnetic field. If you move the magnet, then the magnetic field at any one point in space changes, right? Notice that I'm not saying the strength of the magnet changes; just that if you stood next to a moving magnet, you would notice that the magnetic field strength where you are would change as the magnet moved. If you buy that, then you also have to believe that a changing magnetic field creates an electric field. How do you know that? Because charges in the wire move. If the charges move, that means an electric field is present. So here's an important conclusion:

A changing magnetic field generates an electric field.

We also know from earlier that an electric current generates a magnetic field. Let's rephrase that. Because each moving charge that makes up an electric

---

[19] This explanation section is *not* the same as the corresponding explanation section in the *Stop Faking It!* book on Energy!

current has its own associated electric field, when the charges move that means the electric field at a given point near the charges changes. That leads us to another important conclusion:

A changing electric field generates a magnetic field.

Right about now you might be saying, "So what?" If that's what you're saying, then consider that one model of light is that it's composed of vibrating electric and magnetic fields.[20] If changing electric fields generate magnetic fields and changing magnetic fields generate electric fields, this means that light waves do a good job of keeping themselves going without any other intervention. If that's so, then light waves can keep on moving even in empty space, which is in fact what they do (Figure 4.26).

Figure 4.26

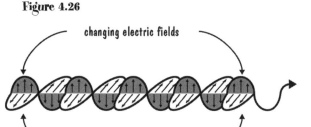

Light waves are composed of changing electric and magnetic fields

A second use for the interaction of electric and magnetic fields is to generate electricity. If you move wire loops in a magnetic field or move magnets around loops of wire, you cause an electric current to flow. Such an arrangement is known as a **generator** (Figure 4.27). The electricity you use in your home or apartment comes from generators that are turned by running water (hydroelectric power), wind, or steam that is created by burning coal, burning oil, or heat from nuclear reactions.

Figure 4.27

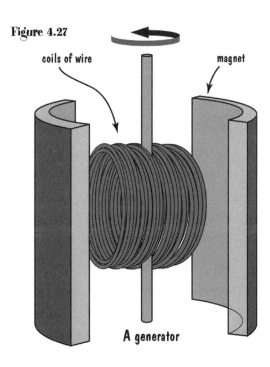

A generator

Topic: Generators

Go to: *www.scilinks.org*

Code: SFEM15

---

[20] See the *Stop Faking It!* book on Light for a more detailed scientific model of light.

Chapter Summary

- Electric currents exert forces on one another.

- Electric currents generate magnetic fields.

- A coil of current-carrying wire behaves like a magnet.

- The presently accepted model of magnetism includes the fact that individual atoms contain *magnetic moments* and act like tiny magnets. The degree to which the magnetic moments in a substance line up determine the degree to which the substance is magnetic.

- External magnetic fields can cause the magnetic moments in a substance to line up. Heat and external forces can cause the magnetic moments in a substance to become jumbled up.

- Moving charges, and therefore electric currents, in the presence of a magnetic field produced by something else feel a force.

- One can derive mathematical expressions for the strength of the magnetic field produced by various configurations of electric current.

- Changing electric fields produce magnetic fields and changing magnetic fields produce electric fields. This principle is behind the generation of electric currents that we use every day.

Applications

1. You can create your own compass without much trouble. Get a sewing needle, a cork, a pan of water, and a bar magnet. Slowly stroke the needle with the magnet, making sure you always use either the north or south pole of the magnet only and that you stroke in the same direction. This causes the magnetic moments in the needle to line up in one direction, so that the needle itself is now a magnet. Then place the needle on top of the cork and place the cork in the pan of water so it floats. The needle will now rotate until it's lined up with the Earth's magnetic field.

2. You already know that the Earth has its own magnetic field, but did you know that the direction of the Earth's magnetic field has changed in the past? In other words, what now is the Earth's north magnetic pole used to be the Earth's south magnetic. How do we know that? We have to combine a bit of Earth science with what we know about magnetism. One of the best accepted models in Earth science is that of plate tectonics, which basically says that large plates in the Earth's crust slowly move across one another and into one another. One part of this model is that there are places in the

ocean where new crust is created and then spreads out in different directions. This is called, strangely enough, "sea floor spreading." If you take samples from the ocean floor leading away from one of the places where the floor is spreading, you will find materials that are magnetized in one direction, then another, as shown in Figure 4.28. The interpretation of these results is that the Earth's crust contains materials that, during formation, have been magnetized by the Earth's magnetic field. Because the direction of magnetization abruptly changes, we infer that the direction of the Earth's magnetic field changed abruptly in the past, and has done so as many as 25 times. And no, we don't understand the mechanism behind these reversals.

Figure 4.28

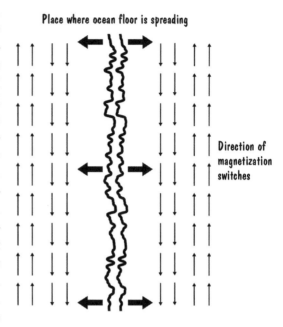

Place where ocean floor is spreading

Direction of magnetization switches

3. Many kinds of metal will become magnetized when you place them in a magnetic field for a long period of time. Screwdrivers often become magnetized and will attract screws when you use them to work on automobiles, because car engines contain strong magnets. A less obvious example is that metal filing cabinets left in one place in a room for a while become magnetized. To see that this is true, find a filing cabinet that's been in the same place for a long time. Place a compass near one of the bottom corners of the cabinet, and slowly move the compass up, along the edge, to the top of the cabinet. The compass will rotate through 180 degrees as you do this, indicating that the filing cabinet is magnetized, with a north and south pole. Why is this the case? Because over a long period of time, the magnetic moments in the metal have lined up with the Earth's magnetic field, leaving the cabinet itself magnetized.

4. The Earth has a natural light show known as the aurora borealis, or northern lights. This light show occurs in the northern latitudes during periods of increased solar activity. Here's what happens. At times the Sun has big ol' storms that send out a stream of charged particles toward the Earth. When these charged particles reach the Earth's magnetic field, they feel a force. The nature of this force (due to that strange thing called a cross product and that trick called the right-hand rule) is that charged particles moving parallel

to the magnetic field lines don't feel any force, and those moving perpendicular to the field lines feel a force that makes them move in circles. The end result is that the charged particles from the Sun spiral in along the Earth's magnetic field lines. Because of the shape of these lines, the particles are funneled in toward the north and south magnetic poles of the Earth. When these particles interact with the Earth's atmosphere, they produce all sorts of pretty colors in the sky. So, we get light shows near both the North and South Poles. Uh yeah, I guess that means there are southern lights as well as northern lights!

5. I mentioned in the previous example that magnetic fields tend to make charged particles move in circular paths. Perhaps you've heard of, or even seen pictures of, things called particle accelerators. What these things do is get subatomic particles moving at very high speeds and then smash them into targets, producing other subatomic particles. Particle accelerators are usually circular in shape, and use strong magnets to keep the charged particles moving along a circular path. After the collisions, the new subatomic particles produced are subjected to a magnetic field. The sizes of the circular paths they trace give information about the charges and the masses of the newly created particles.

Topic: Electromagnets

Go to: *www.scilinks.org*

Code: SFEM16

6. Sometimes it's nice to be able to turn a magnet on and off. One example is when using a magnet on a crane to lift scrap metal from one place and deposit it in another place. To let the scrap metal drop, you need to turn the magnet off. That's easy to do if you use an electromagnet, which is basically what you created with your nail and coil of wire. If you use a coil of wire with an iron center as your magnet, all you have to do is cut off the electricity to turn off the magnet.

7. Electricians use a device called a **galvanometer** (or ammeter) to measure the size of electric currents. A galvanometer is a relatively simple device. The electric current runs through a coil of wire, which is immersed in a magnetic field. This coil of wire is oriented so that when a current runs through it, the resulting forces from the magnetic field cause the coil to rotate. The greater the current, the greater the force and the greater the rotation. Connect the coil to an arrow that indicates the amount of rotation of the coil, and you have something that measures the size of the electric current flowing through the wire.

8. If you've torn a ligament, broken your back, or had any other major injury, chances are you've had an MRI done. MRI stands for Magnetic Resonance Imaging. I can't explain how MRI works completely, because that would

take an extra chapter in the book, but we can understand the basics. First, there's a really strong magnet. That magnet can cause the magnetic moments of hydrogen atoms in various body tissues to line up with its magnetic field. Once they're lined up, these atoms can be stimulated to return to their original state, giving up certain, well-defined bursts of energy in the process. The energy released is where the concept of *resonance* comes in.[21] Detection of these resonances can give a good picture of what the tissues inside your body look like. For example, you can tell whether a tendon or ligament is torn or whether or not a bone is impinging on a nerve, all without cutting someone open to see what's wrong.

9. I told you I'd give you a more accurate model of what happens when a material becomes magnetic, so here we go. Up until now, I've talked about the magnetic moments of individual atoms as if they acted independently of one another. It turns out that these magnetic moments do not act independently, but are in some cases strongly coupled to one another. In fact, the name for the connection is **exchange coupling**, and is an effect associated with something known as quantum mechanics (an involved and downright bizarre model of how things behave on a very small scale such as the scale one encounters when dealing with atoms and their parts). Anyway, materials contain regions, known as **magnetic domains**, in which the magnetic moments of the atoms in the region are perfectly lined up. Figure 4.29 gives an idea of what this might look like on an atomic scale.

Figure 4.29

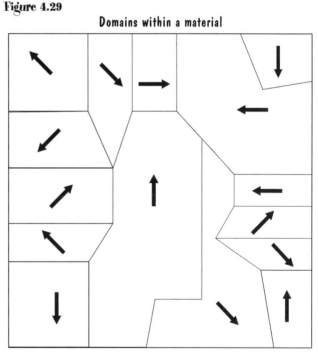

Domains within a material

Magnetic moments within each domain are completely aligned

---

21 See the *Stop Faking It!* books on Light and Sound for a better idea of what the term *resonance* means.

When a material is not magnetized, these domains of various sizes have a random orientation that results in no net effect for the material as a whole. In the presence of an external magnetic field, though, certain magnetic domains in the material grow in size while others decrease in size. If the domains oriented in a certain direction are on average larger than the domains oriented in other directions, the material is magnetic. In permanent magnets, the domains retain their sizes even when the external magnetic field is removed. In other materials, the domains return to their original sizes once the external magnetic field is removed. Basically, this gives the same effect as if the magnetic domains of individual atoms acted independently, but we get the added explanation of why some materials tend to stay magnetic through the coupling of separate atoms.

Cirque du Circuit

This and the next few chapters will be devoted primarily to electric circuits. Understanding the operation of electric circuits and the devices that go into circuits requires a basic understanding of the scientific models of electricity and magnetism, which is why I've saved circuits until the latter part of the book. I want to emphasize the word *understanding*. You can find lots of books and resources that explain how electric circuits work and the rules that govern the construction and operation of electric circuits. As with all the books

Topic: Electric Circuits

Go to: *www.scilinks.org*

Code: SFEM17

in the *Stop Faking It!* series, though, my emphasis is on the reasoning behind the rules. Personally, I had great difficulty in understanding electric circuits when I encountered them in my college physics courses. In retrospect, I had difficulty because I tried to memorize my way through the material. Needless to say, I'll try to spare you that trauma!

If you want to understand electric circuits, then there's no substitute for physically messing around with them. Unfortunately, gathering all the materials you would need can be a bit expensive. Fortunately, we have an alternative, which is the special software that you can download from the NSTA Web site. We'll be able to construct simple to complicated electric circuits using this software, and it's the next best thing to having the actual materials sitting in front of you. Besides, you get to blow up lightbulbs and fry various components without it costing you anything!

Introduction to the software

Before we get into constructing circuits, I want to give a few instructions on using the software. (Instructions for downloading the software are found on page xi in the front of this book.) There's no sense trying to explain all the features here, so I'll just go over the basics. I'll introduce other features as we need them. Anyway, when you open the software you get a screen that looks like Figure 5.1.

Figure 5.1

In the lower left of the screen there's a button labeled "tour." This is an explanation and demonstration of how the software works, so you can always click on this button for a quick review. The largest part of the screen is green, and it's your workspace. You can drag and drop all kinds of electrical components here, hook them up with wires, and see what happens. Speaking of wires, you can create a wire simply by clicking on the workspace where you want it to begin, holding the mouse button down, dragging the wire to where you want it to end, and releasing the mouse button.[1] To change where the wire connects, click on the wire near its end and drag the wire to its new location.

---

[1] I'm referring to the mouse button, rather than the right or left mouse button, because Mac users just have one button on the mouse. For all you PC folks, when I refer to the mouse button, assume I'm talking about the left button.

In the upper left of the screen are buttons labeled "batteries," "switches," "lights," "audio," etc. Clicking on one of these buttons makes objects in that category appear in the window to the right of the buttons. Up and down arrows at the top and bottom of the window let you scroll through the various devices. To choose a device and place it on the workspace, just click and drag the object to where you want it. There's no limit to the number of any one kind of object you can drag to the workspace, so you don't have to worry about running out of anything. When you place the cursor over an object in the vertical window, a description of the object will appear in the horizontal window in the far upper left of the screen. Those descriptions can be useful sometimes.

Just below the workspace are four buttons. "Clear" removes everything from the workspace and "undo" erases your last step. I'll explain what "view charges" and "view schematic" do later. Underneath these four buttons are three windows labeled "Volts," "Amps," and "Ohms." These are meters that measure the magnitude of voltage between two points (I haven't defined voltage for you yet), the magnitude of the electric current running through a wire, and the magnitude of resistance (also not defined yet) between two points.

On the left side of the screen below the window containing devices are four icons. Click on the hand icon to move things around. Click on the circular arrow to rotate devices. Click on the switch icon to turn switches on and off. Click on the wire-cutter icon to remove something from the workspace.

There are eight buttons below the icons. Some are useful and some not. The "labs" button gives you access to a few ready-made circuits that were created for the classroom version of this software. You can play with these circuits if you want, but we won't be using them here. Next down is the "tour" button, which I already explained. Below that is the "Sciclopedia." This contains quick and dirty explanations of various concepts associated with electric circuits, and I wouldn't recommend reading them until you've completed the book in your hands, at which point you'll already know that material! Below that button is one labeled "ideas." Repeatedly clicking on this button produces lots of different cool circuit ideas on your workspace. You should definitely play with these circuits even though I won't address all of them in this book.

Continuing on to the second column of buttons, you have one labeled "open" and another below it labeled "save." These buttons aren't going to do you a lot of good because the save feature is disabled in the software version you've downloaded. Below those two buttons are two labeled "snapshot" and "print." These buttons do work, so if you want to save a graphic file of your work or print a circuit out to hang on the refrigerator, feel free.

Below those eight buttons is an arrow that you can click to view "quick reference cards" that show a few simple circuits and objects. The last button in the lower left allows you to exit the program. Strangely enough, this button is labeled "exit."

Things to do before you read the science stuff

Given that you just read a section on using the software, the next logical step of course is to ignore the software and do something hands on. Don't worry; we'll get to the software soon. Before that, grab your wires and battery you used in the last chapter. You'll also need a flashlightbulb. You can get the bulb by removing it from a flashlight or you can spend a dollar or two and get a new one. Once you have the materials, mess around with them until you are able to use the wires, battery, and bulb to light the bulb. Giving a student these materials and asking him or her to light the bulb is an elementary school science activity that's been around for over 40 years, and it's fun to watch both kids and adults figure this out. I strongly recommend that you try this without

Figure 5.2

looking at the solution, but if you get frustrated, Figure 5.2 shows one way to hook things up. The main trick is in connecting the lightbulb. The connection to one side of the lightbulb filament is on the side of the bulb and the connection to the other side of the filament is on the bottom of the bulb. This is shown in Figure 5.2.

Now that you've lit the lightbulb for real, let's use the software to do the same thing. Drag the 5-ohm, 10-watt, white lightbulb to the workspace. Also drag one of the 1.5-volt batteries to the workspace. Create a couple of wires and hook the lightbulb and battery together as shown in Figure 5.3. Notice that I rotated the battery in Figure 5.3 once I dragged it to the workspace. While you're at it, try the various circuit connections shown in Figure 5.4 to prove that they *don't* work. And if you're really into this, try those various circuit connections with the actual lightbulb and battery to see what happens.

Figure 5.3

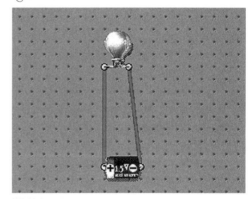

Figure 5.4

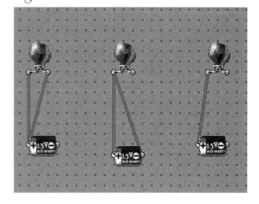

Now add a switch to your circuit, as shown in Figure 5.5. You can use either the wall switch or the single knife switch. Open and close the switch in the circuit, remembering that you need to click on the switch icon in order to operate a switch in the workspace. Once you have your switch working, click on the "view charges" button (just below the workspace) and see what happens when the switch is open and when it's closed. Notice that the charges are negative and notice the direction the charges are moving around the circuit. Click "view charges" again to shut this feature off. Then hold the ctrl button down on your keyboard (the option button on a Mac) and click "view charges" again. This time the charges should be plus signs and they should move in the opposite direction.

Figure 5.5

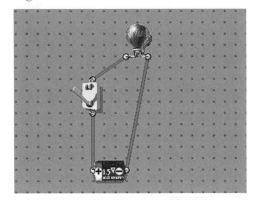

Just for kicks, replace your 1.5-volt battery with the 12-volt battery, keeping the same lightbulb. There's something satisfying about breaking a lightbulb, huh? Of course, this doesn't really happen when a lightbulb burns out, does it?

Before you move on to the explanation section, answer a few questions for yourself. Why does the circuit have to be hooked up the way it is in order for the bulb to light? What's going on inside the wires and inside the lightbulb that creates light? Why is a battery necessary?

The science stuff

When I ask people what's causing a lightbulb to light or a fan to turn or a hair dryer to work, the usual answer is "electricity." All of us have at least a vague idea of some unknown thing called electricity flowing through wires and making things work. In fact, our buddy Ben Franklin thought of electricity as a fluid that moved through metal. Before we get into the details of various electric circuits, I figure I should give you an idea of the currently accepted model of what's happening inside those wires when you hook up a battery and a lightbulb. As we go through the model we'll have a point of confusion, so I might as well address that up-front. The things that move through wires are negatively charged electrons, as you might expect given that in metals, the electrons are essentially free to move all over the place. However, the direction of an elec-

Topic: Electric Current

Go to: *www.scilinks.org*

Code: SFEM18

tric current is defined as the direction that *positive* charges move.[2] So, the direction of electric current is opposite the direction that electrons move. When you click on the "view charges" button in the software, you see negative charges moving around the circuit. They move in the direction that electrons move. Notice that these electrons move away from the negative end of the battery and toward the positive end of the battery, as you might expect negative charges to do. When you hold the ctrl key down (option key on a Mac) and click on "view charges," you see what positive charges would do if in fact positive charges were moving in the wires.[3] This is a useful feature because it shows us the direction of the electric current.

Regardless of the direction of electric current, electrons *do* move around in those wires. Do they speed along so they get from the wall switch in your home to the overhead light in a fraction of a second? Nah. Electrons in a circuit actually move rather slowly, at a speed of about four tenths of a millimeter per second. That translates to moving a foot in about 14 seconds—slow! The reason they move so slowly is that they have lots of atoms to bump into along the way. These collisions generate heat, which is why wires carrying an electric current can get quite hot. You might have noticed that the wire got hot when you created your electromagnet (the nail, wire, and battery) in Chapter 4. Not only can electron collisions with atoms generate heat, they can also generate light. That's a good thing for lightbulbs. Electrons colliding with atoms inside the filament in a lightbulb are what make the bulb give off light.[4]

Right about now you might be wondering how the lights in your home come on immediately when you flick a switch, given that the electrons in the wires don't move very fast. The answer is that the *effect* of turning on the switch gets to the lights quickly, at almost the speed of light. It's like turning on the water to a hose when the hose is already full of water. The water at the end of the hose comes out right away because the "push" from the water at the faucet

---

[2] We can blame Benjamin Franklin for this problem. He had an arbitrary decision to make as to whether positive or negative charges carried electric current. He chose wrong, and we have to suffer for it! You might think that sane physicists would simply change the definition of the direction of electric current, but nooooooo! Tradition rules.

---

[3] Although negatively charged electrons are the things that move in electric circuits, it is possible to generate an electric current with positive charges, as in causing individual protons to move from one place to another. Not a difficult thing to do, as you just have to strip a hydrogen atom of its electron and then use negatively charged plates to send the remaining proton on its way.

---

[4] The scientific model for how this happens is kinda cool, involving electrons jumping from one energy level to another while sending out photons of light. For more details, see the *Stop Faking It!* book on Light.

travels to the end of the hose in a short time. The water that initially comes out of the end of the hose did not travel all the way from the faucet, but was already at the end of the hose when you turned on the faucet. In the same way, the electrons whose collisions cause a lightbulb to light are already in the lightbulb filament when you flip the switch.

Now that we have an idea of what the electrons in the wires are doing, it's time to look at an analogy that will help explain the behavior of all kinds of electric circuits. Imagine you have a series of pipes that contain corn syrup or some other viscous liquid like Prell shampoo.[5] The pipes represent wires and other components in our circuit. Immersed in this liquid are lots and lots of marbles. You might think that the marbles represent electrons, because that's what moves inside electric circuits. Because of our problem with current moving in the opposite direction that electrons move, however, we'll let the marbles represent *positive* charges. Positive charges moving in one direction have the same effect as negative charges moving in the opposite direction, so we'll just blissfully talk about the motion of positive charges, even though we know that it's the negative charges moving around.

Figure 5.6

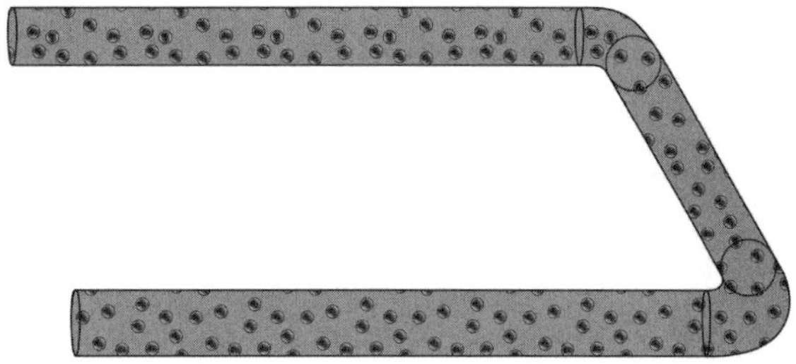

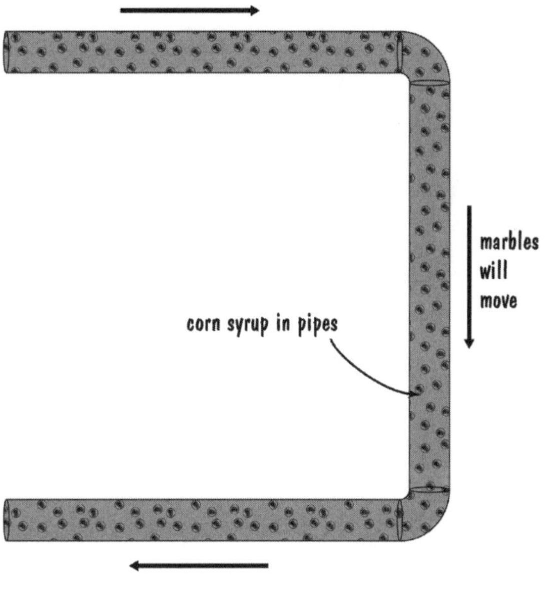

Anyway, Figure 5.6 shows our pipes filled with marbles. Everything is lying flat on the ground, so the marbles just stay where they are.

If we want to get those marbles (positive charges) moving and generate an electric current, we'll have to stand the whole thing upright, as in Figure 5.7.

Figure 5.7

marbles will move

corn syrup in pipes

---

[5] I'm not sure if they still make Prell shampoo, but for those of you who are old enough, I'm sure you remember that commercial that showed a pearl falling through the shampoo bottle. Keep that picture in mind as I go through this analogy.

Figure 5.8

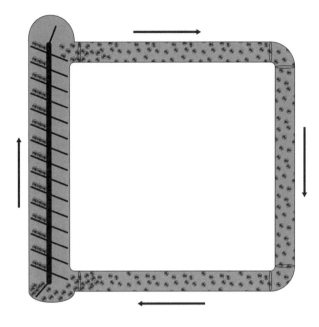

Figure 5.9

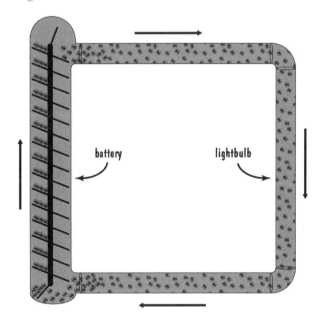

battery   lightbulb

Now the marbles will begin moving, but once they all get to the bottom, there won't be any more motion. If the marbles stop moving, we don't have any more electric current. The solution to this is to create a conveyer belt that takes marbles from the bottom to the top. With that in place, we can keep the marbles moving continuously. Check out Figure 5.8.

Remembering that we're using the marbles to represent electrons that are positively charged, let's see how our model describes a simple circuit. The corn syrup in the pipes keeps the marbles moving at a slow speed, just as real electrons do because of their collisions with atoms. The vertical pipe on the right represents a lightbulb in the circuit. The conveyer belt represents a battery that keeps the electric current going in the circuit (Figure 5.9).

In our analogy, gravity is the force that causes marbles to move naturally from the pipes on top to the pipes on the bottom. We create that situation by standing the pipes up vertically, and we keep the marbles moving by adding a conveyer belt to move marbles from a lower height to a higher height. In an actual electric circuit, electric forces cause the charges to move through the circuit, and we keep those charges moving by using a battery that takes the charges from one place to another. This analogy is so good, in fact, that it's just fine to think of a battery as moving charges through a "height difference," where they can now "drop" back down to the lower height. The proper term to use for this height difference when dealing with electric forces is a **potential difference**, or **difference in potential**.[6] Potential difference is measured in volts. The battery in our circuit actually creates this potential difference. In other words, if you just have a couple of wires and a lightbulb, it's the same as the pipes and marbles lying flat on the ground. Adding a battery stands the pipes up and also provides the conveyer belt to move the marbles from the bottom to the top. Keeping with the gravitational analogy, when the charges move through the lightbulb (marbles drop through the vertical pipe), we refer to that as a *potential drop*. One more vocabulary word before moving on: Anything like a battery that moves charges across a potential difference is known as an **electromotive force**, or **emf** for short.

Okay, let's see if we can use our conveyer belt and pipes to explain the kinds of connections we need in order to light a lightbulb with a battery. Refer back to Figure 5.4 and look at the circuit on the left. There we have the wires hooked up to only one side of the battery. This corresponds to the following setup with the conveyer belt and pipes (Figure 5.10).

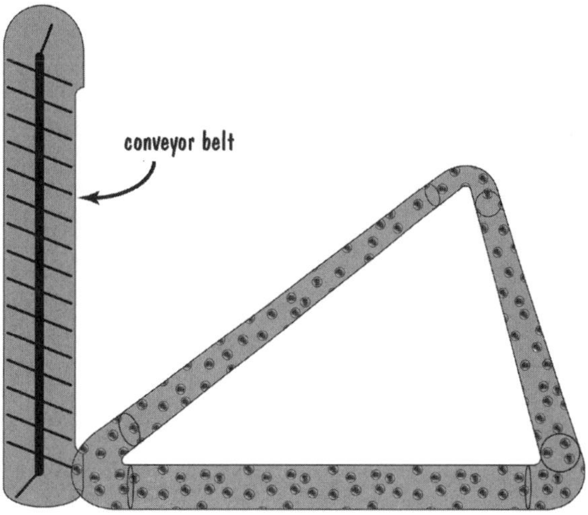

As you can see, this leaves the conveyer belt completely out of the loop. If you have no way of getting the marbles from the bottom to the top, you can't keep the marbles moving. In fact, without a battery to *create* the height difference, the pipes will be on their sides. This corresponds to no electric current, and an unlit bulb. Go ahead and build the first circuit on the left of Figure 5.4. Click on "view charges" to see that no charges are moving.

On to the second circuit in Figure 5.4. Here we're connecting up to only one side of the vertical pipe that represents the lightbulb. Because the con-

Figure 5.10

conveyor belt

---

[6] If you pick up a physics textbook, you'll find a formal definition of *electric potential*, which is related to the concept of electric field lines. I won't go into that definition, because it would probably confuse things rather than make them clearer at this point. You should be fine if you just think of differences in electric potential as differences in height.

veyer belt is hooked up, marbles will definitely move around the pipes, but they won't go through the vertical pipe representing the bulb. No current through the bulb means no light. Construct the center circuit in Figure 5.4 and click on "view charges." You'll see that charges are moving, but not through the lightbulb.

I'll let you figure out on your own why the last circuit on the right in Figure 5.4 doesn't light the bulb, and I'll go ahead and summarize what we need for an electric current to accomplish something for us. First, there has to be a **complete circuit**. That means we have a complete path through which our charges (marbles) can travel. If the path stops at some point, or if there's a break in the path (as with an open switch!), there is no electric current. Second, we need an emf (the conveyer belt) to keep the charges (marbles) moving. Third, if we want an electric current to flow through a device such as a lightbulb, that device has to be a part of the complete path. If we bypass the device, nothing's gonna happen.

SCI**LINKS.**
THE WORLD'S A CLICK AWAY

Topic: Electronic
       Circuits

Go to: *www.scilinks.org*

Code: SFEM19

Before you move on to the next section, make sure you understand how our pipes, corn syrup, and conveyer belt represent an electric circuit. I'm going to be using this analogy quite a bit.

More things to do before you read more science stuff

Before I have you do anything in this section, I'm going to define something for you. Defining a concept before experiencing it goes against what we know about the best way for people to learn things, but the concept is basic enough in this case that the world won't collapse if we violate the principle just this once. Anyway, the concept is **resistance**. As applied to electric circuits, this refers to how much a given device *resists* the flow of electric current. Because electrons bump into atoms even in plain old metal wires, all the wires in a circuit have resistance. Because the filament in a lightbulb resists the flow of electric current more than a wire does, a lightbulb has more resistance than a plain wire. Therefore, in our analogy with pipes and corn syrup, the corn syrup that represents the lightbulb should actually be much thicker than the corn syrup that represents the wires. In fact, the normal thing to do when analyzing circuits is to assume that connecting wires have no resistance at all, even though that's not really true. It's generally a good assumption, though, given that the resistance of wires usually is much less than the resistance of other components in the circuit. By the way, resistance is measured in units of ohms.[7] The resistance of

---

[7] Pretend you're meditating and you'll have the correct pronunciation of these units.

various devices you use in the software program is displayed in the monitor window (upper left of the screen) whenever you place the hand icon over each device.

Now that you know what resistance is, let's construct a few circuits and see what they do. Use the software to set up the circuit shown in Figure 5.11.

The lightbulb in this circuit is the third one down in the collection of lightbulbs, the one that is 5 ohms and 30 watts. Make sure you use that lightbulb. Notice that I've added an ammeter to the circuit. To do that, drag a meter from the bottom window labeled "Amps" and just place it over one of the wires. When everything is hooked up correctly, the "amps" window should read 1.2 amps. For the record, amps is short for **amperes**, which are the units we use to measure electric current. The higher the number, the

Figure 5.11

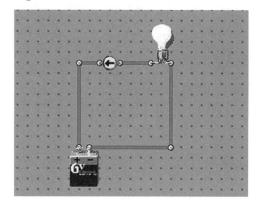

Figure 5.12

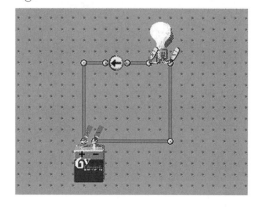

greater the current. Next add voltmeters to the circuit, as shown in Figure 5.12. To do this, you need to use the meters contained in the bottom window labeled "Volts." Drag one probe in that window to the plus side of the battery and the second probe to the minus side of the battery. To access two more probes, just click on a different color in the window. Drag these two new probes to opposite sides of the lightbulb. If you have it set up correctly, the window should show that both sets of probes give you a reading of 6 volts.

Write down the values of the resistance of the lightbulb (5 ohms), the current flowing through the circuit (1.2 amps), and the voltage across the lightbulb (6 volts). Now replace the 6-volt battery with a 1.5-volt battery (you'll have to rotate the 1.5-volt battery and move a wire or two in order to make the circuit work) and note the change in the values of the current going through the circuit and the voltage across the lightbulb. Write those values down. Finally, replace the 1.5-volt battery with a 12-volt battery. Again, write down the values of the current through and voltage across the lightbulb.

5 Chapter

Figure 5.13

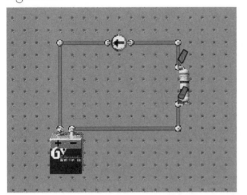

Figure 5.14

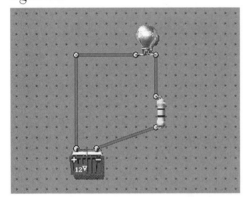

Next click on the "Resistors" button on the left of the screen, so you see a window full of different resistors. All a resistor does is limit the flow of current in a circuit. The larger resistor, the more it slows the electric current. Drag the 10-ohm resistor to the workspace and build the circuit shown in Figure 5.13. Be sure to use a 6-volt battery and to insert an ammeter and two voltage probes in the circuit.

Record the voltage reading and the current reading in the windows at the bottom. Replace the 10-ohm resistor with a 50-ohm resistor and record the readings. Then replace the 50-ohm resistor with a 100-ohm resistor and record the readings.

One more circuit to try. Do you remember when I had you break a lightbulb in the beginning of the chapter? Let's construct that circuit again, but add a 50-ohm resistor to the circuit as shown in Figure 5.14. The lightbulb in this circuit is the top one in the light window, the one you broke earlier.

More science stuff

I told you I'd be using the pipes, corn syrup, and conveyer belt analogy a lot, so let's start with that. For a 6-volt battery connected to a lightbulb, we have the same drawing as in Figure 5.9. The only thing added to this drawing is that the height difference is 6 volts (Figure 5.15).

If we assume that there's no resistance in the wires (the horizontal pipes), then because the conveyer belt raises the marbles up through a height difference of 6 volts, the height drop across

Figure 5.15

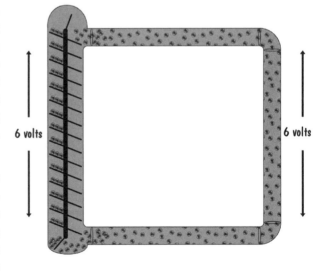

the lightbulb must also be 6 volts. Let's see what happens if we change our battery from 6 volts to 12 volts, but keep the lightbulb the same, with the same resistance. If our height difference is now 12 volts, then our conveyer belt raises the marbles twice as high as before. That means the drop in height across the lightbulb also has to increase to 12 volts, as in Figure 5.16.

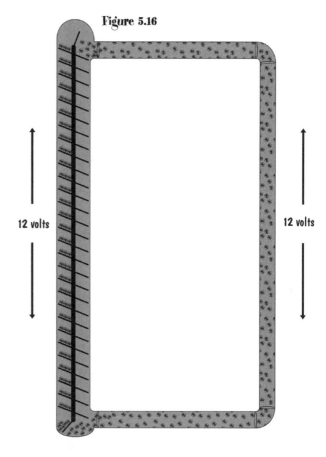

Figure 5.16

12 volts

12 volts

While we're changing the height drop across the lightbulb, its total resistance stays the same as before, which is 5 ohms. In our analogy, increasing the voltage means that the pipe representing the lightbulb has to be longer. If we want that pipe to be longer *while keeping the same total resistance*, that means the corn syrup will have to be thinner than before. Thinner corn syrup means it's easier for marbles to fall through. The more marbles that pass through the pipe in a given time, the larger the current. Therefore, if you increase the voltage while keeping the resistance the same, that should increase the electric current.

Now let's see what happens when you decrease the voltage (use a 1.5-volt battery instead of a 6-volt battery) while keeping the resistance of the lightbulb the same. Now you have a smaller height difference. If the *total resistance of the lightbulb is to remain the same* while *the voltage across the bulb (the height of the pipe) gets smaller,* then you need *thicker* corn syrup in the pipe. That will reduce the number of marbles that pass through the pipe in a given time, leading

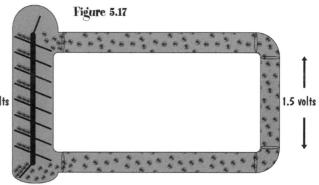

Figure 5.17

1.5 volts

1.5 volts

to a decrease in the current. Figure 5.17 summarizes all of this.

All of this works out mathematically, and we can summarize the relationship between voltage, current, and resistance with the following equation.

Voltage across a device = (current through the device)(resistance of the device)

or

V = IR

where V is the voltage, I is the current, and R is the resistance. This is known as **Ohm's law**, named after German physicist George Ohm, who lived in the nineteenth century. Gee, I wonder why the units of resistance are ohms! To see that Ohm's law works with actual numbers, you can plug in the numbers you recorded for different voltages, currents, and resistance in the basic circuit with a battery and a lightbulb. For example, in the first circuit you built in the previous section (Figure 5.11), you have a voltage of 6 volts, a current of 1.2 amps, and a resistance (the resistance of the lightbulb) of 5 ohms. That leads to

V = IR

6 volts = (1.2 amps)(5 ohms)

Because 1.2 times 5 equals 6, the equation works. You can plug in the numbers you got for the other circuits and get true statements, so it seems that Ohm's law works pretty well.[8] To get a little more insight into the relationship, it helps to picture the equation as a teeter-totter, as in Figure 5.18. The voltage is on the left and the current times the resistance is on the right. Because we have an equation, the teeter-totter is in balance.

Let's keep the resistance in the circuit the same and increase the voltage. That makes the teeter-totter unbalanced to the left, as in Figure 5.19.

Figure 5.18

Figure 5.19

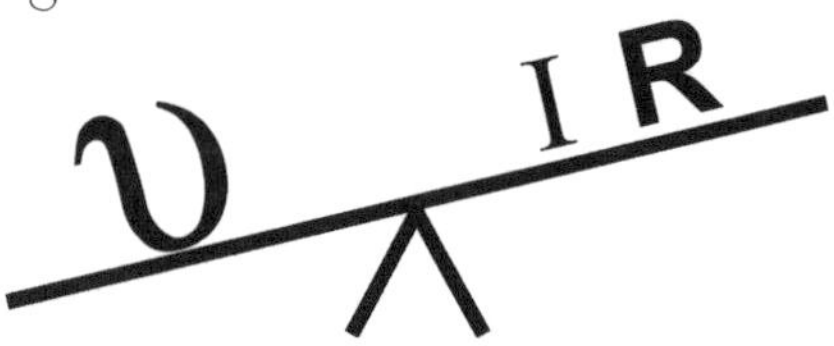

---

[8] For the record, some substances don't obey Ohm's law at all, and virtually all substances have at least a few ranges of current and voltage over which they don't obey Ohm's law. In fact, the tungsten filaments in everyday lightbulbs don't obey Ohm's law completely, as they don't have the same resistance at different currents.

To get things back in balance, *I* must get larger, as in Figure 5.20.

That's the exact same result we got by using our analogy with pipes and corn syrup and such. Increasing the voltage increased the current when the resistance remained the same.

The teeter-totter also tells us what should happen if we keep the voltage the same and increase or decrease the resistance. Increasing the resistance decreases the current, and decreasing the resistance increases the current. Figure 5.21 shows these two cases.

Figure 5.20

Figure 5.21

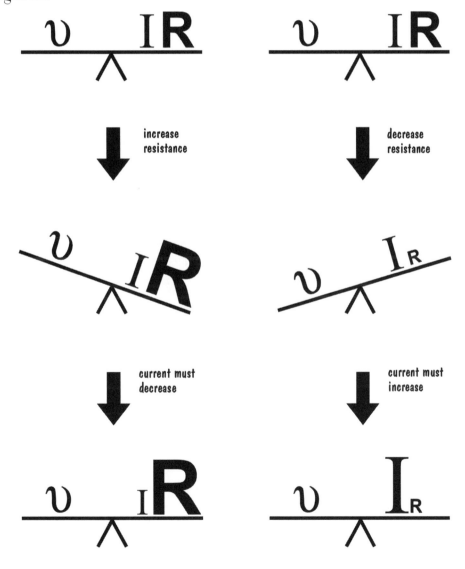

Topic: Current
Electricity

Go to: *www.scilinks.org*

Code: SFEM20

If you're into numbers, you can use the numbers you got for voltage, current, and resistance when you kept the battery the same and changed resistors. $V = IR$ should give you a true statement each time.

You might be wondering why we use resistors in a circuit in the first place. I mean, all they do is alter the amount of current in the circuit. It's not as if they light up or smile at you or play music. The last circuit I had you build in the previous section (Figure 5.14) provides the answer. Without the resistor in that circuit, the lightbulb blows up. With the resistor in the circuit, the lightbulb just lights up. Often we have a fixed voltage to work with, as with the 110 volts you have in your home. Some things you want to plug into the wall would fry if you used all of the electric current that 110 volts can produce, so we use resistors to limit the amount of current flowing through the device. So while resistors might be kind of boring, they perform a valuable service.

Chapter Summary

- For electrical devices to work, you must cause electrons to flow through them.

- Although the *effect* of electron movement can travel rapidly (almost at the speed of light) through an electric circuit, the electrons themselves actually move at quite slow speeds.

- Although negatively charged electrons are the charges moving in virtually all common electric circuits, the direction of electric current is defined as the direction *positive* charges would move—a direction opposite to the motion of electrons.

- For an electric circuit to operate properly, there must be an electromotive force (emf) that drives charges around the circuit, and there must be a complete path through which the charges can travel.

- Sources of emf create *potential differences*, which drive electrons around a circuit. When electrons travel through a device that has resistance, those electrons are said to have undergone a *potential drop*.

- The resistance of an object, measured in ohms, determines the extent to which the object limits the flow of current in a circuit.

- Voltmeters measure the potential difference, in volts, between any two points of a circuit.

- Ammeters measure the electric current, in amps, flowing through any portion of a circuit.

- The voltage across and the current through any device or series of devices are related by Ohm's law, which is $V = IR$. Most devices obey Ohm's law within at least some range of values of current and voltage.

Applications

1. You can use your actual battery and lightbulb circuit and your virtual battery and lightbulb circuit to find out which materials conduct electricity and which don't. Use tape to connect your battery, bulb, and wires as shown in Figure 5.22. Then put the free ends of the wires on two parts of any object such as a coin, a stick, and a refrigerator. If the lightbulb lights, your object conducts electricity. If the lightbulb doesn't light, your object doesn't conduct electricity and is called an **insulator**.

Topic: Insulators

Go to: *www.scilinks.org*

Code: SFEM21

Though it's a bit less exciting, you can do this same thing with the software. Set up the circuit shown in Figure 5.23 and place various objects from the "Other" window on the left between the open ends of the wires. Some objects might conduct electricity but have so much resistance that they reduce the current enough that the lightbulb doesn't light, so use "view charges" to see whether or not there is a current. Use the 12-volt battery when you check the pickle—it might just surprise you!

Figure 5.22

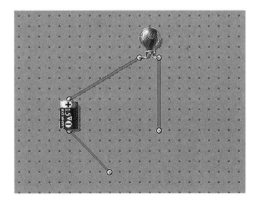

2. You undoubtedly know that many people die each year from electrocution in their homes. Many other people get a healthy shock without serious damage. When I was a kid, there was a lady who helped my mom clean our house once a week. When cleaning, she would unscrew the lightbulb in the oven and clean the socket with a wet rag. I'm sure she got a few shocks (her hair looked that way), but she never died while I was around. Why is it one person can be stu-

Figure 5.23

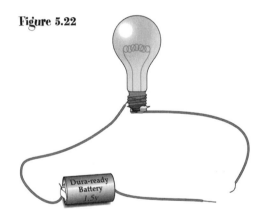

pid and stick a knife in a plugged-in toaster and just get a jolt, while someone else can die? The answer has to do with resistance. The normal resistance of human skin is pretty high, which means it does a good job of keeping the current that might flow through someone's body relatively small. Shoes generally have an even higher resistance than human skin, so it's difficult to create a well-conducting electric circuit from an electric outlet through your body to the ground. You can change all that by being even more stupid than washing a lightbulb socket or sticking a knife into a toaster. If you don't wear shoes, the resistance between you and the ground decreases. Because water is a good conductor of electricity, standing in water reduces that resistance even more.[9] Having your skin wet makes *you* a better conductor of electricity and increases your chances of being electrocuted. So, it's not a good idea to take a bath or a shower around electric appliances. Current building codes require electric outlets that are near sinks or bathtubs to have what's known as a *ground fault interrupt*. This is basically a fuse (more on fuses in the next chapter) that shuts off the electric current if a person becomes intimately involved with an electric current headed from the outlet to the ground.

3. Ever wonder why different electrical devices require different-sized batteries? Some require a single 1.5-volt battery, others require four 1.5-volt batteries, and others require a 9- or 12-volt battery. To see the reason for different sized batteries, open the software and drag either the fan or the color spinner from the "Other" window to the workspace. Then drag different-voltage batteries, from 1.5 to 12 volts, to the workspace. Successively connect each battery to the fan or color spinner and notice the difference in speeds. In general, more electric current means more speed. So, we use different voltage batteries for different devices simply because the different devices require different electric currents to operate properly.

---

[9] Pure distilled water actually is a poor conductor of electricity. However, most of the water we use every day has lots of dissolved minerals that make the water a good conductor.

Cutting, Splitting, and Stacking Circuits

We've dealt with how basic circuits work, but there's more to using electricity than hooking up a battery and lighting a lightbulb. Electrical devices you use every day are a bit more complicated. For example, how can they make strings of Christmas tree lights that don't all go out when one bulb burns out? What's going on with those circuit breakers that switch off when you operate too many appliances at once? Answers to these and other questions as we proceed.

Things to do before you read the science stuff

Set up the two circuits shown in Figure 6.1.[1] Notice the brightness of the bulbs in each circuit and see if you can explain what's going on.

Put an ammeter in each circuit (get them from the Amps window) to measure the current in each circuit. Also measure the voltage across each lightbulb using voltage probes from the Volts window. When measuring the current and voltage, the placement of your ammeters and voltage probes should look something like Figure 6.2. Because there are only three sets of voltage probes, Figure 6.2 only shows three possible placements of the probes. Feel free to move the probes around to check the voltage between any two points in the circuit. As you do this, note that each pair of probes is color-coded and that the readings in the window reflect the voltage between two probes of the same color.

As long as you're explaining why the bulbs are as bright as they are, go ahead and explain why the currents and voltages are what they are. Can't explain it? Don't fret, as I'll be providing the explanation in the next section. Use the clippers icon to remove one of the bulbs from the circuit on the right in either Figure 6.1 or Figure 6.2. What happens to the other bulbs in the circuit?

Figure 6.1

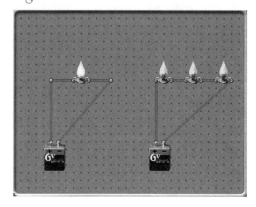

Figure 6.2

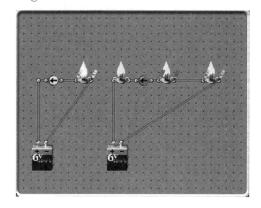

SC/LINKS.
THE WORLD'S A CLICK AWAY

Topic: Electric Power
Go to: *www.scilinks.org*
Code: SFEM22

---

[1] As with all the circuits I ask you to build, I will assume you're using the software. If you have access to enough wires, batteries, and lightbulbs, then by all means build these circuits for real.

Next create the two circuits shown in Figure 6.3. Again notice the brightness of the bulbs. Notice that instead of there being a single path through which electric charges can flow, the wires split up into three separate paths, one for each bulb, and then come back together.

You already know what the voltage and current readings are for the circuit on the left, because that circuit is identical to the one on the left in Figures 6.1 and 6.2. So, check out the current and voltage readings in various parts of the circuit on the right in Figure 6.3. In Figure 6.4 I've shown a few possible placements of the ammeters and voltage probes, but again because of the shortage of meters, I haven't shown all possibilities. Feel free to move the ammeters and voltage probes to different parts of the circuit. While you're at it, check the voltage between two points on a single section of wire. You should get a reading of zero volts. Does that surprise you?

Using the clippers icon, remove one of the bulbs from the right-hand circuit in Figure 6.3. What happens to the other bulbs? What happens if you remove two of the bulbs? Can you figure out why this happens?

While you're pondering explanations for what has happened so far, let me introduce you to a new meter in the software, known as an ohmmeter. An ohmmeter measures the resistance between two points. To see how the ohmmeter works, connect each of the three meters available in the Ohms window to three different devices, as in Figure 6.5.

You don't have to choose the same things I've chosen, but you should notice that the readings on the ohmmeters match the descrip-

Figure 6.3

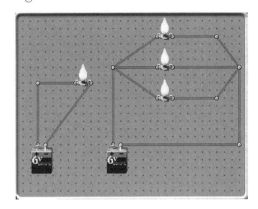

Figure 6.4

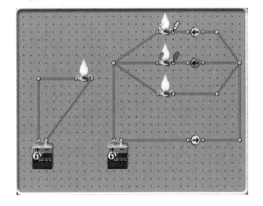

Figure 6.5

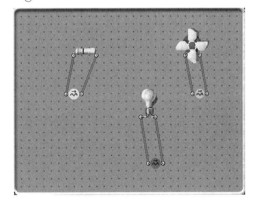

Figure 6.6

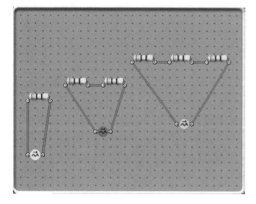

Figure 6.7

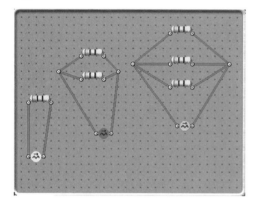

Figure 6.8

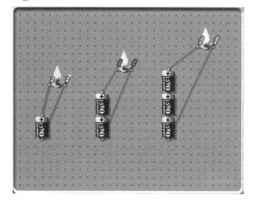

tions of the devices in the window in the upper left of the screen. In case you're wondering, yes, ohmmeters do rely on Ohm's law to figure out the resistance of something.

Next measure the resistance of various combinations of resistors, as I have done in Figure 6.6. Those are all 10-ohm resistors. So your numbers will match with mine in the explanation section, do me a favor and use 10-ohm resistors yourself. Write down the ohmmeter readings for each combination.

Figure 6.7 shows combinations of resistors (still using 10-ohm resistors) that are a bit different from the combinations shown in Figure 6.6. In the second and third combinations, notice that the wires split up into separate paths rather than being along one path as in Figure 6.6.

Check out the readings on the ohmmeters and answer the following questions. How can *adding* resistors result in a combination that has *less* resistance than using just one resistor? Shouldn't adding resistors result in greater resistance?

We've been stacking up resistors and lightbulbs, so why not stack up a few batteries? Construct the circuits shown in Figure 6.8, with voltage probes placed across each lightbulb. Notice that I have connected the batteries directly to one another, without any wire in between. That's allowable, but you can also put wires in between if you want. Note the difference in brightness of the lightbulbs in the three circuits. Maybe you can even explain why the bulbs are as bright as they are, and why the voltage probes read what they do. If you're really on top of things, you can figure out what the current in each circuit should be and then check your answer using an ammeter (you'll have to rearrange the wires in the circuits so you have at least one vertical or one horizontal wire in which to place the ammeter).

In Figure 6.9, I've switched the direction of one battery in the two circuits on the right in Figure 6.8. Compare the voltages across the lightbulbs and the brightness of the lightbulbs in Figure 6.9 with those of Figure 6.8.

Before getting to the explanations, let's try one more combination of batteries. Set up the three circuits illustrated in Figure 6.10. Measure the voltages across the lightbulbs and the current in each circuit. Is there any advantage to using extra batteries in this kind of combination?

The science stuff

Before getting into explanations of what you've observed, I want to make sure one thing is clear. The purpose of wires in the software, and for practical purposes in actual circuits, is just to connect various devices together. Even though we know that

Figure 6.9

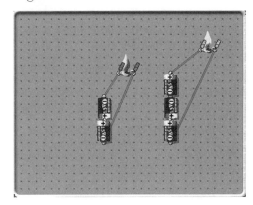

Figure 6.10

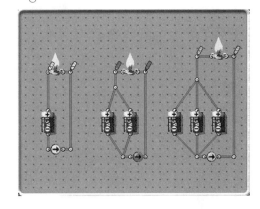

all wires have *some* resistance, we ignore that resistance when figuring out what happens in electric circuits. The software we're using actually assigns zero resistance to all wires. In real life, we use excellent conductors of electricity, such as copper, for wires. That generally ensures that the resistance of the wires is negligible in comparison with other components of a circuit.

Okay, let's move on to understanding all the circuits I had you build in the previous section. In Figures 6.1 and 6.2, there is only one possible path for the electric current to take. The electric current has to go through the battery and through all of the lightbulbs in the circuit. That means one lightbulb in the circuit on the left and three lightbulbs in the circuit on the right. Any electric circuit that has only one possible path the current can follow is called a **series circuit**. To understand what's going on in a series circuit, we can use the analogy of pipes and such. We already know what that analogy looks like for the series on the left in Figures 6.1 and 6.2. Figure 5.9 shows that analogy, with the battery being represented by a conveyer belt and the lightbulb being represented by a vertical pipe with extra-thick corn syrup in it. The analogy for the circuit with three lightbulbs is similar in that there is only one battery (conveyer belt). There are, however, three lightbulbs. Thus, we need three lightbulbs that "share" the voltage (height) drop that was assigned entirely to one lightbulb when only one lightbulb was in the circuit. Figure 6.11 shows this.

Figure 6.11

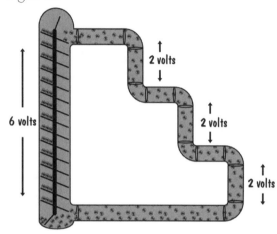

Because each of the lightbulbs is responsible for only a third of the total voltage drop, there is a drop of only 2 volts across each bulb. This smaller voltage leads to dimmer bulbs. A dimmer bulb also means there must be less current going through the bulbs than when there's only one bulb in the circuit. Your ammeter should have told you that, with 1.2 amperes flowing through the circuit with one bulb and 0.4 amperes flowing through the circuit with three bulbs. More on why the current is smaller a bit later in this section.

When you remove one of the components in a series circuit, everything stops working. The reason is pretty simple. When a component is removed, you no longer have a complete path along which the charges can flow. No current; nothing works. So, remove one lightbulb in the right-hand circuit and all the other bulbs go out.

Figure 6.12

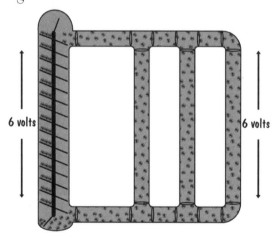

Moving on to the right hand circuit in Figures 6.3 and 6.4, notice that there are three possible paths for the electric current to take. Circuits that split up into separate paths that then join back together are called **parallel circuits**.

I'm sure you noticed that each lightbulb was just as bright as the lightbulb in the circuit on the left, and that the voltage across each lightbulb was 6 volts rather than 2 volts. Let's head to our analogy in Figure 6.12 to figure out why. This time, instead of the three lightbulbs sharing the 6-volt height difference, the circuit splits into three paths, each path containing a single bulb or corn syrup–containing pipe. Each pipe drops the entire height difference of 6 volts.

That corresponds to a voltage reading of 6 volts across each bulb, which you should have verified with your voltage probes.

So, with 6 volts across each lightbulb, it makes sense that each bulb will be as bright as a single bulb connected to a 6-volt battery. If you checked the electric current through each separate path, as I asked you to do, you should have found that the current in each path was 1.2 amps, just as with a single bulb and the 6-volt battery. If you checked the current in the wire that connects directly to the battery, you found that there were 3.6 amperes flowing in that wire. What that means is that you don't get something for nothing. You get three lightbulbs that are just as bright as a single lightbulb connected to the battery, but the battery has to produce three times the current in order for that to happen.[2]

If you remove one of the lightbulbs in this arrangement, the other lightbulbs remain lit. The reason is that, although removing the lightbulb from one path cuts off electric current to that path, the other paths remain intact and able to carry an electric current. This is a characteristic of all parallel circuits—when you remove one component from the circuit, the components that are in the re-maining paths continue to operate.

To help all of this make even more sense, let's look at the measurements you made of the resistance of various combinations of resistors. When you combine resistors in a single path as in Figure 6.6, that's called *combining the resistors in series*. When combining resistors in series, the total resistance is just the sum of the resistances of the separate components. One 10-ohm resistor gives you 10 ohms of resistance. Two 10-ohm resistors give you 20 ohms of resistance, and three 10-ohm resistors give you 30 ohms of resistance. This way of adding resistors is almost common sense, because using the pipes and marbles analogy tells us that the total resistance has to be the sum of the individual resistances—the marbles have to pass through each pipe containing corn syrup (see Figure 6.13).

Figure 6.13

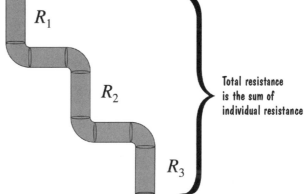

Total resistance is the sum of individual resistance

---

[2] Any time you have a physical situation in which it appears that you *are* getting something for nothing (in other words, you get more energy out than you put in), chances are your analysis of the situation isn't correct. Either that, or much of what we know about physics is wrong!

Because I know you love formulas, here's one that describes the resistance when you add resistors in series:

$$R_{total} = R_1 + R_2 + R_3 + ...$$

where R_{total} is the resistance of the combination,[3] and the terms on the right side are the resistances of the individual components. The three dots on the end indicate that you can add as many resistors as you want, and it's still just a matter of adding up the individual resistances. That's a pretty simple formula, because all you do is add things up. Just so you don't think all formulas are that simple, let's move on to adding resistors in a different way.

In Figure 6.7, you added resistors in such a way that each resistor was in a different section of wire. This is known as *adding resistors in parallel*. If you were to send an electric current through this combination, the current would split up into two or three different paths. Now, when you measured the resistance of these combinations, you should have found that two resistors gave you 5 ohms and three resistors gave you 3.3 ohms. In other words, adding resistors *lowers* the total resistance. Why is the resistance lower when you add resistors? The key is that in adding resistors, you also are creating extra paths. Creating extra paths makes it *easier* for electric charges to flow, even if the extra paths contain really large resistors. To see that, imagine that you have just graduated from high school and are applying to colleges. Just before you apply, you discover that a new university is opening, one that has the most stringent admission requirements that ever existed. Would your reaction be, "Oh no! They've added a new, tough university. Now it will be harder than ever to get into college!" Well, if that's your reaction, maybe you should stay in high school a while longer. Adding a new university, even one that's really difficult to get into, makes it *easier* overall to get into college. With more colleges available, your chances of getting into a college are better than before. In the same way, adding a new path in which electricity can flow, even a path with lots of resistance, makes it easier for electric current to get through the combination. Therefore, the total resistance of resistors combined in parallel is *less than* the resistance of any single resistor. Why yes, we do have a formula that allows you to calculate exactly the resistance of resistors combined in parallel. Here it is:

$$\frac{1}{R_{total}} = \frac{1}{R_1} + \frac{1}{R_2} + \frac{1}{_3R} + ...$$

As with the previous formula, R_{total} stands for the total resistance of the combination and the other Rs are the resistances of the individual resistors. I'll explain

---

[3] In most textbooks you will see the total resistance referred to as the *equivalent resistance,* with the symbol being R_{eq}.

later where this formula comes from. For now, let's see if it matches what you've observed so far. Let's add two 10-ohm resistors in parallel, as pictured in the center of Figure 6.7. Substituting into the above formula gives us

$$\frac{1}{R_{total}} = \frac{1}{R_1} + \frac{1}{R_2}$$

$$\frac{1}{R_{total}} = \frac{1}{10} + \frac{1}{10}$$

$$\frac{1}{R_{total}} = \frac{2}{10}$$

$$\frac{1}{R_{total}} = \frac{1}{5}$$

This means that R_{total} is equal to 5, which is what you measured with the ohmmeter. If you've forgotten how to add fractions and that last little bit freaked you out, don't worry; I won't be testing you on that process. If adding fractions isn't a big deal for you, you can go through the calculation for three 10-ohm resistors and find that R_{total} is equal to 3.33. Otherwise, just take my word for it.

Just for kicks, let's go back to Figures 6.2 and 6.4 and try to make sense of the current and voltage readings you got there. In the right side of Figure 6.2, you have three lightbulbs connected in series. Each lightbulb has a resistance of 5 ohms. Because these are in series, we simply add up the individual resistances to get the total resistance. 5 + 5 + 5 is 15 (news flash!), so the total resistance is 15 ohms. Now we just use Ohm's law to figure out the current flowing in the circuit.

$$V = IR$$

To solve for the current, I, we divide both sides of this equation by R. That leaves us with

$$\frac{V}{R} = I \quad \text{or} \quad I = \frac{V}{R}$$

Now we just plug in 6 volts for V and 15 ohms (the total resistance) for R.

$$I = \frac{6 \text{ volts}}{15 \text{ ohms}} = 0.4 \text{ amps}$$

That's the reading you got on the ammeter in the first place. Next, let's apply Ohm's law to the circuit on the right in Figure 6.4. There the lightbulbs are in parallel, so we add their resistances using the formula for adding up resistors in parallel.

$$\frac{1}{R_{total}} = \frac{1}{R_1} + \frac{1}{R_2} + \frac{1}{R_3}$$

$$\frac{1}{R_{total}} = \frac{1}{5} + \frac{1}{5} + \frac{1}{5}$$

$$\frac{1}{R_{total}} = \frac{3}{5}$$

That means that R_{total} is equal to $\frac{5}{3}$. To find the electric current in the main wires of the circuit (remember, we're treating that combination of resistors as one resistor), we use Ohm's law again.

$$\frac{V}{R} = I$$

or

$$I = \frac{V}{R}$$

$$I = \frac{6 \text{ volts}}{5/3 \text{ ohms}} = 3.6 \text{ amps}$$

Again, this is the reading you got when using the ammeter in this circuit.

All right, enough math! All we have left to explain is what you get when combining batteries in series and in parallel. In Figure 6.8, you added batteries by stacking them on top of one another, with all of them pointing in the same direction (minus on bottom and plus on top). This is just like stacking conveyer belts on top of one another, as in Figure 6.14.

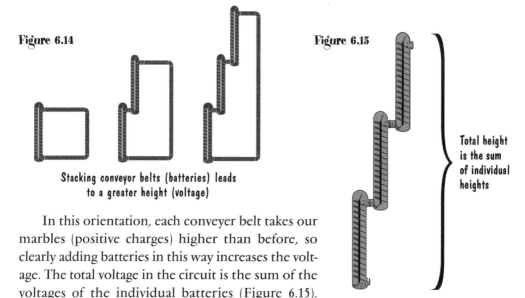

Figure 6.14

Stacking conveyor belts (batteries) leads to a greater height (voltage)

Figure 6.15

Total height is the sum of individual heights

In this orientation, each conveyer belt takes our marbles (positive charges) higher than before, so clearly adding batteries in this way increases the voltage. The total voltage in the circuit is the sum of the voltages of the individual batteries (Figure 6.15).

That's why the light gets brighter as you move to the right in Figure 6.8.

So, if you want to combine batteries to get a greater voltage, you simply connect them together plus to minus. Now do you know why it's important to pay attention to the plus and minus signs when putting batteries in a CD player or a camera? If not, check out what happens when even one of your batteries is positioned in the wrong direction. That's what I had you do in Figure 6.9, and it corresponds to inserting a "down" conveyer belt in our analogy. That reduces the overall voltage, which is generally not a good thing (see Figure 6.16).

Finally, what happens when you add batteries in parallel, as in Figure 6.10? Basically nothing, as we obviously don't increase the voltage by doing that (the lightbulb doesn't get any brighter as you move to the right in Figure 6.10). Figure 6.17 illustrates this using our friendly analogy.

So, is there any advantage to connecting batteries in parallel? Yes, but not much of one. If you check the currents in the wires in Figure 6.10, you'll find that with batteries connected in parallel, each battery has to supply less current. That increases the life of each battery. Of course if you happen to be someplace where it's difficult or expensive to replace batteries (say Antarctica), longer battery life could be a good thing.

Time for a recap. In general, there are two types of electric circuit—series circuits and parallel circuits. Of course, complicated circuits can be a combination of both series and parallel circuits. We can use relatively simple rules to figure out the total resistance of various combinations of components, which in turns tells us the currents in the wires of a circuit. Batteries can be added in both series and parallel, but the combination that's most useful, providing the batteries are lined up in the proper direction, is a se-

Figure 6.16

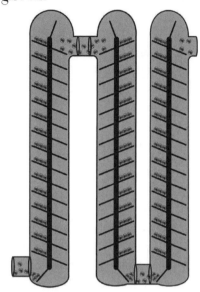

A battery in the wrong direction is like a "down" conveyor belt

Figure 6.17

Parallel conveyor belts don't increase the height

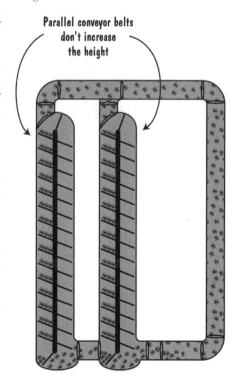

ries combination. The advantage of parallel circuits over series circuits is that in a parallel circuit, the rest of the components still operate when one component fails. The disadvantage of parallel circuits is that they require more total current from a battery or other emf.

SCLINKS.
THE WORLD'S A CLICK AWAY

Topic: Batteries
Go to: *www.scilinks.org*
Code: SFEM23

More things to do before you read more science stuff

Use the software to build the circuit illustrated in Figure 6.18. Take your time and realize that the components don't have to be in exactly the same places they are in the figure, and that the purpose of the wires is to connect things together, so the wires can be any length or at any angle as long as you make the necessary connections. The lightbulb on top is a 10-watt bulb and the two lower lightbulbs are 30-watt bulbs, but since I don't refer to specific numbers below, it really doesn't matter what lightbulbs you use. Make sure you place ammeters in the circuit (you'll have to rotate two of them) where indicated. Also note that I've clicked on ctrl-view charges (option-view charges on a Mac) so the electric *current* is shown and that the ammeters point in the direction positive charges move (as opposed to the direction electrons move).

If you wanted, you could figure out what's going on in this circuit by figuring out which components combine in series and which in parallel. It might take a while, but you could do it with what you already know. We're not going down that road, though. Instead, I want you to concentrate on the junction where I've placed the ammeters. Current enters that junction from the top and left, and leaves through the wire heading straight down. Check out the readings on those ammeters and see if you can discover a relationship between the three currents.

Next relocate the upper ammeter to a different wire, as shown in Figure 6.19.

The three ammeters now outline a lower loop in the total circuit. That loop contains the two batteries, one lightbulb, and the spin-

Figure 6.18

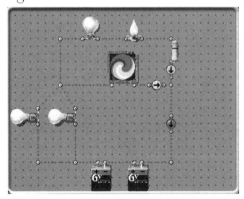

Figure 6.19

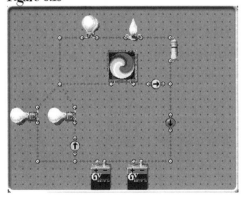

ning color wheel. Use voltage probes to measure the voltage across the lightbulb (the one on the right that's in the loop we're focusing on) and the color wheel. You already know the voltage across the two batteries, given that they're 6-volt batteries.

More science stuff

Before explaining, let me make sure you got the results I expected you to in measuring the currents and voltages. Within an error of about a thousandth of an amp, you should have found that the current readings in the wires to the top and left of the specified junction added up to the current in the wire leading down from that junction. As for the voltages in the lower loop, I hope you found that the voltage across the lightbulb plus the voltage across the color wheel added up to 12 volts. These readings illustrate a couple of important principles in circuit analysis, and as usual I'll use pipes, marbles, etc., to explain them.

Figure 6.20 shows the analogy for the entire circuit illustrated in Figure 6.18. Take a few moments to convince yourself that Figure 6.20 is a good representation of Figure 6.18 (would I lead you wrong?).

Figure 6.20

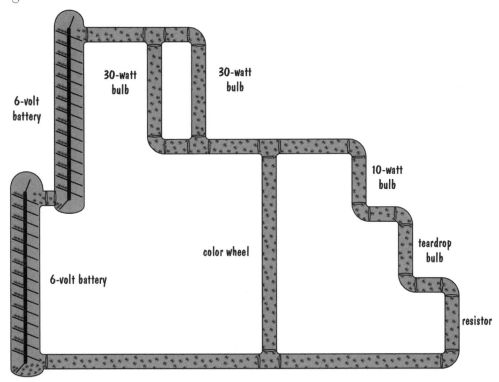

First concentrate on the junction with the dotted line around it (Figure 6.21).

Figure 6.21

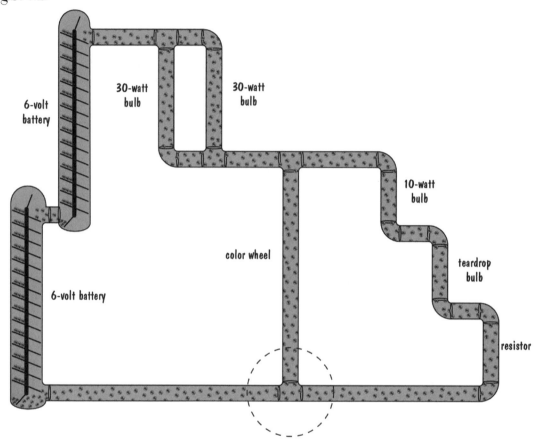

Here you have marbles (current) coming from the color wheel and the resistor (pipes with corn syrup), both leading into the pipe that goes to the negative side of one of the 6-volt batteries (the bottom conveyer belt). Would you believe that the number of marbles per second entering this junction from the color wheel and resistor equal the number of marbles per second leaving the junction and heading to the battery? Of course you would. If that didn't happen, it would mean marbles were disappearing into thin air or appearing out of nothing. Well, if you believe that marbles don't just appear and disappear, then you should believe the following:

**The sum of electric currents entering a circuit junction equals
the sum of electric currents leaving that junction.**

Pretty much common sense, huh? Okay, if you buy that one, then try this one.

The sum of the voltage rises and drops around any closed loop in an electric circuit is equal to zero.

That might not be so obvious, so let's see how it applies to our analogy. Figure 6.22 shows that analogy with attention called to the loop in which I had you measure voltages.

Figure 6.22

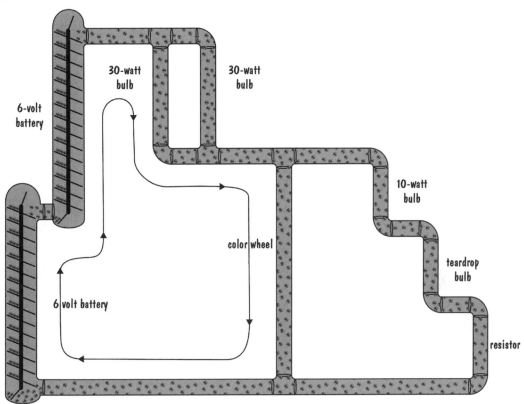

Start at any point in this loop. Add up all the voltage rises (where you go upwards in the figure) and voltage drops (where you go downwards in the figure) and you should end up back where you started. In other words, those rises and drops add to zero. Again, common sense. Of course, if our analogy holds true, then the voltage rises you measured in the electric circuit (6 volts plus 6 volts for the batteries) should equal the voltage drops you measured (the voltages across the lightbulb and the color wheel). Now according to our rule, this should hold true for *any* loop you choose in the circuit. Figure 6.23 shows most of the possible loops you can take. If you want to check out the rule, then measure the voltages across all the other components in Figure 6.18, trace a path around any one of the loops, and see that the total voltage rises equal the total voltage drops. As you do this, keep in mind that you don't need batteries to have a voltage rise. All you have to do is go "against the current" across any of the elements to achieve a voltage rise.

Figure 6.23

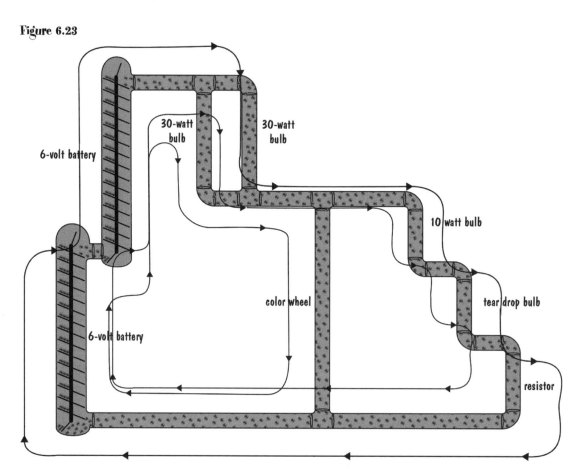

The two rules stated in boldface above are known as **Kirchhoff's rules**, named after Prussian-born mathematician and physicist Gustav Kirchhoff. These rules are quite useful for figuring out what currents will flow in a circuit and what voltages will be across all the components in the circuit, no matter how complicated the circuit. A nice bit of information to have if you want to ensure that everything in your circuit has enough current to operate properly but not so much current that you fry the components.

I promised earlier that I would explain the formula for finding the total resistance of resistors added in parallel, namely

$$\frac{1}{R_{total}} = \frac{1}{R_1} + \frac{1}{R_2} + \frac{1}{R_3} + ...$$

I'm now ready to do that. Take a look at the circuit shown in Figure 6.24.

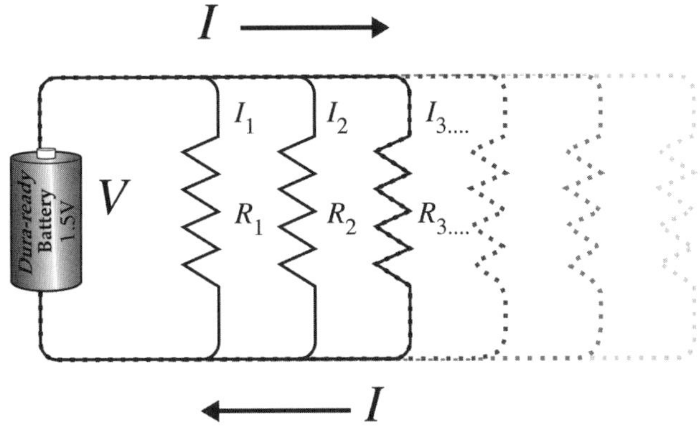

There's a battery with a voltage V and a bunch of resistors that are added in parallel, with resistances R_1, R_2, R_3, etc. The electric current that runs through the battery is I, and this current splits up into currents I_1, I_2, I_3, etc. Because the current I separates out into all those other currents, we know from the first of Kirchhoff's rules that I equals the sum of all those smaller currents, or

Figure 6.24

$$I = I_1 + I_2 + I_3 + ...$$

We also know that the voltage across the entire combination of resistors is the same as the voltage across each individual resistor. To see that, all you have to do is construct the pipes and marbles analogy and realize that the battery raises the marbles up a height V and the marbles drop through a height V no matter which path they take. Finally, we can use Ohm's law to come up with a different expression for each of the currents listed in the equation above. We can replace the current through each resistor with the voltage across the resistor divided by the resistance itself.[4] In doing that, we end up with

$$I = I_1 + I_2 + I_3 + ...$$

$$\frac{V}{R_{total}} = \frac{V}{R_1} + \frac{V}{R_2} + \frac{V}{R_3} + ...$$

Then, through the magic of algebra, we can divide both sides of this equation by V, leaving us with

$$\frac{1}{R_{total}} = \frac{1}{R_1} + \frac{1}{R_2} + \frac{1}{R_3} + ...$$

---

[4] If this doesn't make sense to you, head back to the first half of this chapter and find where I started with $V = IR$ and ended up with $I = \frac{V}{R}$.

which is the formula for adding resistances in parallel. If you didn't quite follow the algebra, don't sweat it. Plans are in the works for a *Stop Faking It!* book on mathematics!

Even more things to do *plus* even more science stuff

So far in this chapter, we've done a fair amount of splitting and stacking, but not much in the way of cutting. So, it's time to introduce you to various switches and fuses, and their purposes in circuits. You'll also use the "View Schematic" button, which should be really exciting. I'm going to break from tradition a bit and mix up the activities and explaining in this section, simply because that will work best for this subject matter.

For starters, set up a simple circuit that contains a battery, a bulb, and a knife switch, as shown in Figure 6.25.

Figure 6.25

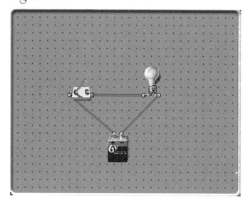

Click on the "View Charges" button and close the switch (remember that you have to click on the wall switch icon labeled "Switch Mode" before closing the switch). No big surprise here. Closing the knife switch completes the circuit and lights the bulb.

Replace the knife switch with a wall switch. Same results, right? One difference, though, is that you can't really see what's going on inside the wall switch. To correct that situation, click on the "View Schematic" button. When you click on this button, all of the electric devices are shown as they're represented when an electrician or scientist draws them on paper. Each device has its own symbol, and the symbols are relatively easy to draw, such as a jagged line for a resistor and a series of short and long lines for a battery. Makes life simple for those of us who aren't artists. Anyway, the advantage of the schematic view for our purposes is that it reveals what's actually happening inside a device. In the case of the wall switch, the schematic view shows that it's basically the same as a knife switch, opening and closing the circuit.

Grab the green push-button switch from the switches window at the left and put it in place of the wall switch. Push and hold the button and the light stays lit. Let go of the button and the light goes off. Click on the "View Schematic" button to see how the push-button switch works. When you push the button you complete the circuit, and when you release it the circuit is open. Okay, so what? Doesn't seem all that useful, does it? Ah, but it is. For example, all of the keys on your computer keyboard and the buttons on your computer mouse are push buttons. Hold them down and the circuit is complete, causing a certain action, such as typing the letter dddddddddddddd, to continue. Release the button and the circuit is broken, causing the action to stop. Push buttons are also great for doorbells. Replace the light in your circuit with the doorbell that you'll find in the "audio" window to the left. Now when you push the button and let go, you get a familiar "ding dong." Push the button and hold it in and all you get is the "ding." Let the button go and you get the "dong." How does this doorbell work? To find out, click on "View Schematic." You'll see that the doorbell contains a coil of wire inside, which we know from Chapter 4 is an electromagnet. When you turn the electromagnet on by pushing the button switch, the piece of metal inside the coil of wire moves to the right. When you release the button, the metal moves to the left. This action causes the ringer to alternately give you a ding and then a dong.[5]

Set up your basic battery and bulb circuit, and insert a thermal switch (you can find this in the switches window) in the circuit. If everything is hooked up correctly, the light should blink on and off. Look at the thermal switch and you'll see that it alternately opens and closes. How is it doing that, you say? Well, a thermal switch is composed of something called a **bimetallic strip**, which is a long strip composed of two different kinds of metal. When an electric current flows through a bimetallic strip, the metals, like all other conductors, heat up.[6] This causes the two metals to expand at different rates, and the strip bends, opening the circuit.[7] After the circuit is open, the strip cools and straightens out again, closing the circuit. Then the strip heats up again and opens the circuit. Obviously, this will continue as long as there's an emf in the circuit. Gee, do you suppose the blinking hazard lights on your car use a thermal switch?

---

[5] The particular arrangement inside the doorbell included in this software is not the simplest possible arrangement, and in fact it would take me about a page to explain exactly how it works. Here are some key ideas, though. The metal rod inside the coil only moves *as* you turn on the current and *as* you shut it off—in other words, when the current in the coil of wire is *changing*. When the current in the coil is changing, the associated magnetic field is changing. Changing magnetic fields generate electric fields, which in turn have their own magnetic fields. Suffice to say that a changing current in the coil causes the metal rod inside to act like a repelling magnet for a very short time, a time long enough for the clapper above to hit a chime.

---

[6] In case you've forgotten, conductors heat up because of the collisions between electrons and the atoms in the conductor.

---

[7] For a more complete explanation of bimetallic strips, see the *Stop Faking It!* book on Energy.

Okay, on to the next kind of switch, which is called a **relay switch**. Grab a relay switch (not the *double* relay switch) from the switches window and build the circuit pictured in Figure 6.26 around it. The lightbulb in this circuit is the 30-watt bulb—the third one down in the lights window.

Figure 6.26

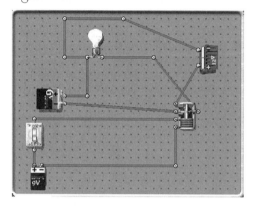

Yes, that is a pretty complicated circuit compared to what we've dealt with so far, but it's not so difficult to figure out if you take it slowly. Make sure your connections to the relay switch are in the right places before you continue. To see what this circuit does, flip the wall switch on and off, and notice that the brightness of the lightbulb changes.

To begin understanding this, click on the "View Schematic" button and just concentrate on the relay switch and the lower, rectangular path that contains the 9-volt battery and the wall switch. When you turn the wall switch on and off, the relay switch changes from one contact to another. Before investigating the usefulness of such a switch, let's figure out how the relay switch works. In the schematic view, you can see that the lower two contacts on the relay switch are connected to a coil of wire, which is just an electromagnet. Turning the wall switch on runs an electric current through this electromagnet, causing it to be, well, a magnet.

At this point, the schematic drawing of the relay switch in our software is a bit misleading, so I'll give you a drawing that's more accurate, as in Figure 6.27.

When you turn on the electromagnet inside the relay switch, it pulls down on the right side of the strip of metal inside the switch (magnets attract some metals, remember?). That causes the left side of the metal strip to move up and touch the top contact on the relay switch. When you turn off the elec-

Figure 6.27

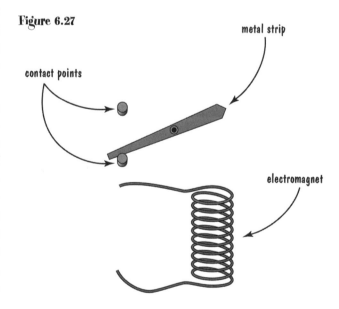

metal strip

contact points

electromagnet

tromagnet, the metal strip returns to its original position, touching the lower contact on the relay switch. This process is shown in Figure 6.28.

Now let's look at the other wiring in this entire circuit. Click whatever buttons are necessary so you can see moving charges and can see everything in a schematic view. With the wall switch off, you'll notice that the only path that has any current flowing in it is the one that contains the 6-volt battery and the lightbulb. Therefore, we're operating the bulb with a 6-volt battery. When you flip the wall switch to the on position, current flows in two different

Figure 6.28

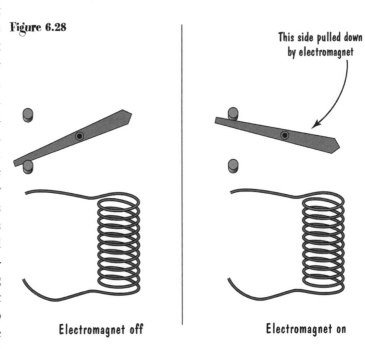

This side pulled down by electromagnet

Electromagnet off   Electromagnet on

paths. One is the lower, rectangular path (this is the current that causes the relay switch to change positions). The other is the path that contains the 12-volt battery and the lightbulb. The overall result is that you now have a 12-volt battery lighting the lightbulb. Of what use is this? How about switching from low beams to high beams on automobile headlights? There are other uses for relay switches, such as having a warning light go on if a current is or isn't flowing in a specified circuit. You can click on the "Ideas" button in the software to see several examples of the uses of relay switches.

Speaking of the "Ideas" button, click on it once. You'll see a bunch of Christmas lights and what is known as a distributor switch. Study this circuit for a while and you should see that the purpose of a distributor switch is to alternately allow current to flow not in just two circuits, but many different circuits. As long as we're on an automobile theme, the distributor in a car does pretty much what the distributor switch in this circuit does, alternately firing spark plugs in different cylinders of the car engine.

Enough with switches, whose main purpose is to turn circuits on and off. Let's move on to devices whose main purpose is to turn circuits off and leave them off.

Build the circuit shown in Figure 6.29. You can find the heater in the "Other" window. Notice that all the batteries are rotated on their side. To avoid having to rotate each battery individually, you can rotate the battery that's in the batteries window and then just drag and drop a bunch of rotated batteries to the workspace.

Figure 6.29

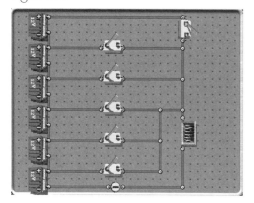

Notice that there's an ammeter in the very bottom wire, so you can see how much current is flowing through the heater. So you can easily see what's happening when you close various switches, it will help to click on the "View Charges" button.

Okay, all set up? Close the bottommost switch. The heater should operate, the ammeter should read 2.4 amps, and charges should be flowing in the bottommost loop. If that doesn't happen, check to see that all the wires actually connect to something and aren't just near where they should be. In other words, check to see that you have a complete circuit. Next open the bottom switch and close the next one up. Note the current flowing through the heater. Then open that switch and close the third switch from the bottom. Do you see that the higher up the switch you use, the more batteries get into the picture and the larger the current through the heater?

Keep closing higher and higher switches, making sure that as you do that, you open all the switches below the one that's closed. Note the current flowing through the heater each time. Something should happen when you get to the topmost switch, which is actually on the top right, namely that the heater gets zapped. Of course, with real heaters you don't keep connecting up 12-volt batteries until the heater blows, but things can happen, such as the crossing of wires or the connection of too many appliances, that increase the

Figure 6.30

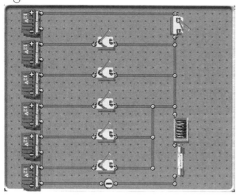

current to dangerous levels. Let's suppose that this is a real heater, and that instead of getting zapped it starts a fire. That wouldn't be a good thing. What would be great is if we had a way to shut off the current to the heater before it got too large to cause problems. Well, we have devices called **fuses** that do just that. Go to the "breakers" window and drag the top fuse in that window to the workspace. Place it in the circuit as shown in Figure 6.30 (make sure all the switches are open first).

Close the bottommost switch. What will

happen is that the fuse will fry, creating an open circuit, but the heater will remain unscathed. Well, we didn't let the current get too high and we saved the heater, but we don't have any heat. If you look at the upper left window with the cursor over the fuse you used, you'll notice that it's a 1-amp fuse. That means it fries whenever there's a current of 1 amp or larger through it. So, all we have to do is get a fuse that has a higher rating, such as 10 amps. Scroll down the "breaker" window until you find a 10-amp fuse and put it in the circuit as in Figure 6.31.

Now try increasing the voltage by closing switches, starting at the bottom and making sure you open lower switches as you close higher ones. Now you should be able to operate the heater until the current gets to 10 amps or above.

Figure 6.31

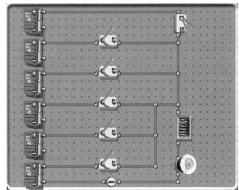

Okay, great. We now have a method for protecting appliances and keeping electric currents to a safe level. Lots of things, including cars and expensive electrical equipment, use fuses for just this purpose. The only problem is that if you blow a fuse you have to replace it. That can be inconvenient and expensive. The solution to this problem is to use something known as a **circuit breaker**. A circuit breaker acts just like a fuse in that it creates an open circuit when the current gets too high, but it has the advantage that it doesn't destroy the breaker. To complete the circuit again, all you have to do is flip the switch on the circuit breaker. Choose the 10-amp circuit breaker from the "Breakers" window and put it in place of the fuse in our circuit. Notice that when the current gets above 10 amps the circuit breaker opens the circuit. When you open the top two switches and cause a lower current to flow in the circuit, switch the circuit breaker to the on position and everything works just fine. Older homes have a fuse box that protects the various circuits in the home. Newer homes have a "breaker box" that contains nothing but circuit breakers. Same function but more convenient.

Figure 6.32

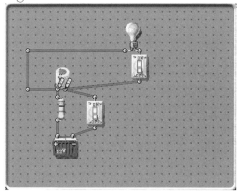

One last neat device to cover in this chapter. Get a radio that operates using an AC outlet adaptor. If the radio also uses batteries, remove the batteries. With the radio plugged in, turn on the radio for a bit and then turn it off. Next unplug the radio. Next turn on the radio with it still unplugged. It should work for a few seconds. How? To give you a clue, create the circuit shown in Figure 6.32.

You undoubtedly recognize the battery, the wall switches, the lightbulb, and the resistor. That cylinder-shaped object is called a **capacitor**, which you will find, of course, in the "Capacitors" window to the left of the workspace. Click on "View Schematic," and the capacitor will look simply like two plates with a gap between them—an open circuit.

Notice that I've place two voltage probes across the capacitor. Click on "View Charges" and then turn the lower wall switch to the on position. Watch the reading on the voltmeter after you turn the switch on. It should slowly increase to a bit over 11 volts. Turn the lower wall switch off and take another look at the voltage reading across the capacitor. Obviously, you have somehow "stored" a potential difference across the capacitor.

Leaving the lower switch off, turn on the upper wall switch. Watch what happens to the lightbulb and watch what happens to the voltage reading across the capacitor.

So, capacitors can do two things. First, they can store a potential difference by maintaining a separation of positive and negative charges. Second, they can release that stored potential difference quickly or slowly. If they release that stored energy slowly, they can create cool effects such as causing a lightbulb to dim slowly (Christmas lights, anyone?). If they release the energy quickly, they can help provide a burst of energy that might be useful in some circumstances. One example of that is the flashbulb in a camera. The small batteries in the camera don't provide enough power to provide a bright flash, so the camera uses capacitors to store up energy that can be released all at once.

Topic: Battery

Go to: *www.scilinks.org*

Code: SFEM24

On to the radio. Most radios use capacitors to store energy. When you turn on an unplugged radio that had been previously plugged in, the energy stored in the radio's capacitors gives you sound for just a bit. Radios also use capacitors to tune to different radio stations, but understanding that application is beyond what we can understand at this point.

Chapter Summary

- Any complete circuit or portion of a circuit that has only one possible path through which electric charges can flow is known as a *series circuit*.

- Any complete circuit or portion of a circuit that has more than one possible path through which electric charges can flow is known as a *parallel circuit*.

- Most circuits are a combination of parallel and series circuits.

- In a series circuit, the voltage generated by the emf in the circuit is shared among the components of the circuit.

- In a parallel circuit, each component sees the full voltage generated by the emf in the circuit.

- The total resistance of devices added in series in a circuit is equal to the sum of the resistances of the individual devices.

- The total resistance of devices added in parallel in a circuit is less than the resistance of each individual device, and is given by $\dfrac{1}{R_{total}} = \dfrac{1}{R_1} + \dfrac{1}{R_2} + \dfrac{1}{_3R} + ...$

- Batteries and other emfs combined in series in a circuit add or subtract their voltages, depending on their orientation.

- Kirchhoff's rules describe the behavior of currents and voltage rises and drops in an electric circuit. Those rules are as follows: *The sum of electric currents entering a circuit junction equals the sum of electric currents leaving that junction. The sum of the voltage rises and drops around any closed loop in an electric circuit is equal to zero.*

- There are a number of different switches that one can use in electric circuits. Some require a physical action and others require a current to operate.

- Fuses and circuit breakers protect circuits and the devices in them by cutting off the flow of electricity when the current gets too high.

Applications

1. There are few things more annoying than having one light in a string of Christmas lights go out, causing all of the lights not to work. To find out which bulb is bad, you have to go through and check each bulb individually. At least that's the way things used to be, when Christmas lights were all connected in series. Modern Christmas lights are connected in parallel, or in a combination of series and parallel. That way, when one light goes out the rest remain lit, or at least only a portion of the lights in the string go out.

2. By now you probably have guessed that much of the wiring in your home or apartment is composed of parallel circuits. If everything were connected in series, turning off a light in the living room would cause the refrigerator to stop working! Of course, wiring a home can be a complicated task. You have appliances that require different voltages and currents, and you can't have too many appliances on one circuit or the current will get too high and blow a circuit breaker. This is where Kirchhoff's rules come in. You can use these laws to plan out what goes where in the various circuits. You can also use these laws to figure out how much resistance to add internally to various devices to make sure they operate properly.

3. Do you have any dimmer switches where you live? If not, you have undoubtedly run across a dimmer switch somewhere—they're great for that "just right" lighting when the kids are gone. Wait, the kids are never gone. Oh well, dimmer switches are still cool. A dimmer switch is nothing but a variable resistor. To see one in action, choose a variable resistor from the resistors window and place it in a circuit with a battery and lightbulb. When you turn the knob on the variable resistor, that changes how much resistance it provides in the circuit (use the schematic view to see what's going on inside the variable resistor). Changing the resistance changes how much current is in the circuit and changes how bright the bulb is.

4. Voltmeters and ammeters both use a galvanometer (a meter that deflects when there's a current in it), but in different ways. In either case, you want to measure the desired quantity (voltage or current) without disrupting the circuit. Figure 6.33 shows the setup for both a voltmeter and an ammeter.

The voltmeter connects to the circuit by providing a parallel path. Because the voltmeter shouldn't draw much current from the circuit, it contains a really large resistor. The very tiny current that goes through the voltmeter is enough to determine the voltage between two points (using Ohm's law) without doing much to affect the main circuit.

Figure 6.33

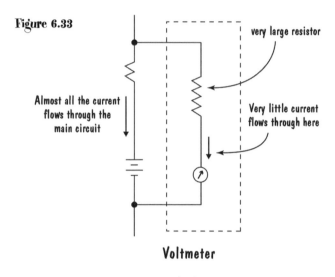

Voltmeter

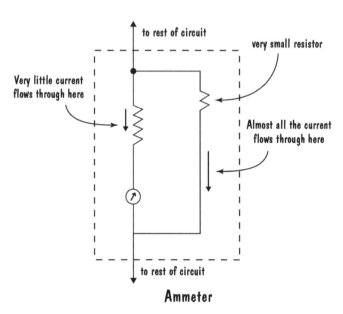

Ammeter

An ammeter uses two resistors. The resistor that's in series with the meter is very large, so little current goes in that path. Parallel to that path is a wire with a very small resistor, so virtually all of the current goes through that path. Again, you have a small current going through the meter without doing much to affect the rest of the circuit.

5. Most appliances come with a three-pronged plug these days. What's with that third prong? It's supposed to make the appliance safer. Does it? Yep. Figure 6.34 shows a drawing of an appliance and the wires that lead from a wall socket to the appliance.

Figure 6.34

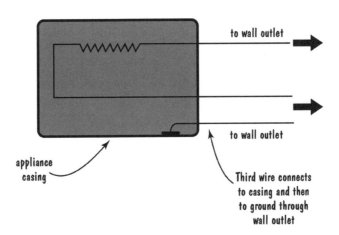

to wall outlet

to wall outlet

appliance casing

Third wire connects to casing and then to ground through wall outlet

Two of the wires operate the appliance. These are the wires connected to the two prongs in a two-pronged plug. Suppose something goes wrong inside the appliance and one of these wires accidentally touches the metal casing on the appliance. Now you have a bad situation. If a person touches the casing, that completes a circuit from one connection to the wall socket, through the person, and onto the ground. Possible electrocution. The third wire eliminates this possibility. Because the third wire is connected to the casing on the appliance and also to the ground (by way of the wall outlet), any accidental contact between one of the other wires and the casing completes a circuit from the casing directly to the ground. To make this situation even safer, you can install a circuit breaker in the wall outlet that blows whenever there's a direct connection between the third wire and ground. This circuit breaker is called a ground fault interrupt, and it's required by most housing codes whenever an outlet is near a sink, bathtub, or shower.

6. If you're ever stranded with a dead battery and need to use jumper cables to start your car, there's a good way and a bad way to connect the jumper

cables. The proper way is to connect the positive terminal of the good battery to the positive terminal of the bad battery, and negative to negative. The reason for this is obvious if you draw the associated circuit. Figure 6.35 shows the proper connection and the improper connection.

Figure 6.35

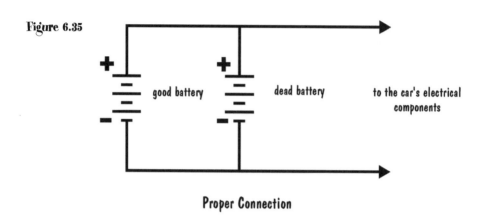

Proper Connection

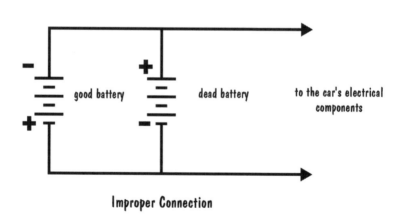

Improper Connection

In the proper connection, the batteries are connected in parallel, with no problem. Both batteries are trying to drive the current through the circuits of the car in the same direction. In the improper connection, the good battery simply drives the current through the bad battery, with most of the current staying in this loop rather than going through the car's circuits. Lots of current through a car battery can cause it to blow up. That's really not very good, especially given that car batteries contain sulfuric acid.

7. Feeling pretty good about your knowledge of electric circuits? If so, try to figure out the following circuit: Many houses have combinations of switches, either for hallways or staircases, where you can turn a light on and off from either of two locations. That means you can turn a light on at one location and then turn it off at another location. To test your prowess, try to figure out a simple circuit that will accomplish this task. In order to do that, you'll need a switch we haven't yet used, called a "two-way single switch." You can find that switch in the switches window, and you need to use two of them. If you're not feeling particularly confident, simply take a look at the solution (Figure 6.36) on the next page.

Figure 6.36

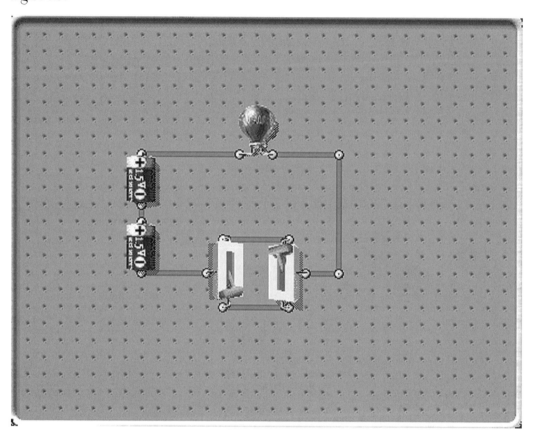

Direct from High Voltage to You and Your Computer

We've covered most of the basics of electricity, magnetism, and electric circuits, but there are still a few things we haven't touched on. For example, what's up with high-voltage power lines, why is working on them a more dangerous job than changing a lightbulb, and how in the heck do computers do what they do? Stay tuned for those answers and more.

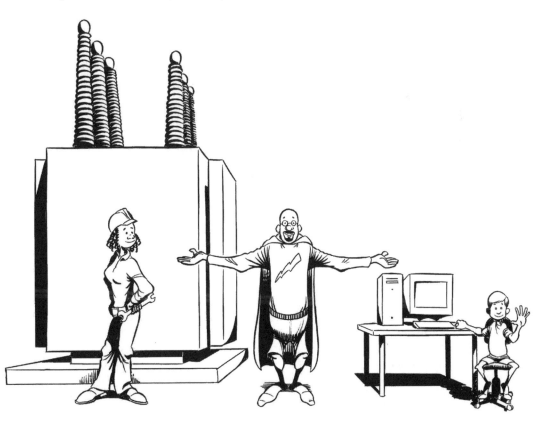

Things to do before you read the science stuff

For this section, I'm just going to have you think about a few things and try to answer a few questions.

1. You might have heard that electric current comes in two forms, AC and DC. What do those letters stand for and what's the difference between the two? How do you produce these two kinds of electric current?

2. You've undoubtedly seen public service messages regarding how dangerous high-voltage power lines are. If they're so dangerous, why do we not just use low-voltage power lines? What happens between those high-voltage lines and your house that makes the electricity you use somehow less dangerous?

3. Why can we plug some things like lamps directly into wall sockets, while other things like computers and train sets and cell phones and electric razors require adaptors?

The science stuff

1. *You might have heard that electric current comes in two forms, AC and DC. What do those letters stand for and what's the difference between the two? How do you produce these two kinds of electric current?*

Besides bringing to mind a marginal heavy metal band,[1] the letters AC and DC stand for **alternating current** and **direct current**. Direct current is what we've been dealing with so far. Direct current flows only in one direction in a given wire. Alternating current, as you might expect, changes direction.[2] Direct current is what batteries give you, and alternating current is what's available from wall sockets.

---

[1] The musical opinions of the author do not necessarily reflect the views of the management of NSTA Press (feel free to substitute your own legal-type statement), but I'd put money on not being the only person who works with this fine group of people who shares this opinion!

---

[2] Mathematical analysis of alternating current is just something I'm not going to mess with at this point, given that the procedures go well beyond basic algebra. One thing you should know, though, is that it's customary to use small letters (such as *i* and *v*) to describe alternating current. I won't bother using small letters for AC in this book, but don't be surprised to find them used in most science books. Also, there is a quantity known as **impedance** in alternating current that one uses instead of the concept of resistance. Spend five minutes talking to a sound-system salesperson and you'll undoubtedly hear low impedance as being good, high impedance as being bad, and impedance matching as being essential. Think "resistance" every time you hear "impedance" and you'll probably be nodding your head at the correct times!

The way to generate direct current is to get electrons where they wouldn't tend to go all by themselves. In a battery, that means getting them to the negative terminal of the battery so they'll head through wires and devices in order to get to the positive terminal of the battery. In the analogy I've been using, this is like the conveyer belt getting marbles where they wouldn't tend to go all by themselves (against the pull of the Earth's gravity) so they can do what they naturally do, which is fall down under the influence of Earth's gravity. I won't get into much detail regarding how a battery gets electrons where they don't tend to go, but I can tell you that it's accomplished through chemical reactions. As I discussed in Chapter 1, different atoms and molecules have different affinities for electrons. All you have to do is find two different substances that tend to exchange electrons in such a way as to create a potential difference between two points, and you have a battery.[3] Because chemical reactions tend to proceed in one direction only (they're one-way conveyer belts), these chemical reactions produce direct current. You can also produce direct current without chemical reactions, and I'll discuss how after explaining how we get alternating current.

In Chapter 4, I discussed one device you can use to generate an electric current called a **generator**. Generators operate on the principle that charges moving in a magnetic field feel a force. Spinning a bunch of wire in a magnetic field causes the electrons in the wire to move, giving you an electric current. Figure 7.1 shows a drawing of a generator.

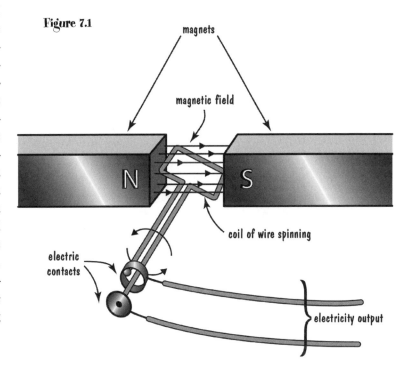

Figure 7.1

magnets

magnetic field

N

S

coil of wire spinning

electric contacts

electricity output

---

[3] The fact that you need two separate substances to get a battery is reflected in the common names for batteries, such as carbon-zinc and nickel-cadmium.

SCⁱLINKS.
THE WORLD'S A CLICK AWAY

Topic: Magnetic Fields
of Power Lines

Go to: *www.scilinks.org*

Code: SFEM25

Topic: Electrical Energy

Go to: *www.scilinks.org*

Code: SFEM26

Because of the nature of the force that magnets exert on moving charges, wires spinning inside a magnetic field will feel forces that cause them to move first in one direction and then in the opposite direction. The faster the wires spin, the more often the electrons switch directions. How often the electrons switch directions is known as the **frequency** of the current, measured in cycles per second. The alternating current that comes from a wall socket has a frequency of 60 cycles per second.[4]

One can also use a generator to create direct current, but it takes clever connections in the generator to accomplish that. There's a reason for sticking to alternating current, though, as I'll explain in answering the next question.

2. *You've undoubtedly seen public service messages regarding how dangerous high-voltage power lines are. If they're so dangerous, why do we not just use low-voltage power lines? What happens between those high-voltage lines and your house that makes the electricity you use somehow less dangerous?*

The reason we use high-voltage power lines has a lot to do with the concept of **energy** as it applies to the conduction of electricity, so we need to discuss that first.[5] Hopefully by now you have a mental picture of some kind of emf (a battery or generator)[6] giving energy to electrons by virtue of their motion and position. The energy these electrons have shows up in various useful and not-so-useful ways. The units we use to measure energy are joules. What's of primary interest in dealing with electricity is not so much the amount of energy that we generate and use, but the *rate* at which we generate and use that energy. The rate at which something gains or loses energy, in joules per second, is known as **power**. We measure power in units of **watts**. That ought to be a familiar term to anyone who's used a hair dryer, bought a stereo system, or bought and/or

---

[4] The frequency of alternating current in Europe is 50 cycles per second, which tends to create problems for people traveling in Europe and expecting to use things like CD players without also using something to convert the 50 cycle per second current to 60 cycles per second.

[5] This will be an *extremely* brief discussion of energy—just enough to get us where we need to be at this point. For a much more thorough discussion of energy and power, see the *Stop Faking It!* book on Energy.

[6] In case you've forgotten, emf stands for *electromagnetic force*. An emf is any device that gets electrons or other mobile charges to a position or configuration they would *not* tend to get to all by themselves. In our gravitational analogy, an emf is represented by a conveyer belt.

screwed in a lightbulb. We all know that a 100-watt lightbulb is brighter than a 75-watt bulb, which is brighter than a 60-watt bulb. Makes sense, because the higher the wattage on a bulb, the greater the rate at which the bulb dissipates energy as light and heat.

I haven't really given you the proper background for deriving a math expression that tells how much power is generated in an electric circuit (that would take another chapter), so I'll just state that expression. It's

$$P = IV$$

where P is the power, I is the current through a given part of a circuit, and V is the voltage across that part of the circuit. So, if you want to know how much power you're generating, you simply measure the voltage and current at the place you're generating the electricity. That might be a hydroelectric dam or some other source of energy.

Now, it would be really great if all the power generated at a power plant eventually made it to various homes and businesses, but that's not the case. We have to transmit that power over transmission lines (basically just wire), and we know what happens when you send an electric current through wire—the electrons bump into atoms and generate heat. That heat takes away energy that otherwise could be used constructively. Obviously, then, we'd like to minimize the heat lost as we transmit electricity.

Because transmission lines are resistors, we know that Ohm's law, or $V = IR$, applies. If we substitute IR in for V in our expression for power above, we get

$$Power = (Current)(Voltage)$$
$$P = IV$$
$$P = I(IR)$$
$$P = I^2R$$

The term I^2R is known as **joule heating** because it tells us how much power is lost due to the heat generated in a resistor. Because the current is squared in this expression, how much current is flowing through a resistor (in this case we're talking about a transmission line) strongly affects the heat lost in that resistor. That's reasonable, because the larger the current, the more electrons per second are flowing and bumping into other atoms, and the bumping is what dissipates energy. So as we transmit electricity along transmission lines, we want the current to be as small as possible. How can we do that? To find out, we have to go back and look at our original expression for power, $P = IV$. This power that you generate is the same regardless of the values of I and V, because you are putting a set amount of energy into the system each second. If P is constant, we know how to get the current, I, to be small. Just make V as large as possible. Maybe another teeter-totter will make that clear, as in Figure 7.2.

After all that, we have an answer as to why we transmit electricity at high voltages. By making the voltage large, we make the current small. If the current is small, we lose as little power as possible to joule heating in the transmission wires.

Okay fine. We're transmitting electricity at high voltage and losing as little energy as possible to heat. Well, not fine, actually. High voltage power lines carry electricity at hundreds of thousands of volts, which is really dangerous! Not the kind of voltage people want in their homes or even at the sites where they generate the electricity. Fortunately, there's an easy way to transform high-voltage power to low-voltage power, and vice versa. We use what's called a **transformer** (see—some terms in science actually make sense!). Take a look at Figure 7.3. There we have two coils of wire next to each other.

Figure 7.2

Because P = IV, the teeter-totter is in balance.

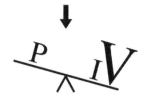

Making V much larger unbalances the teeter-totter. Hence P is not equal to IV.

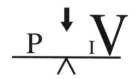

Because P is constant, I must get much smaller to put the thing back in balance.

Figure 7.3

Figure 7.4

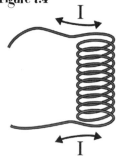

Run an alternating current through this coil. That creates a changing magnetic field.

The changing magnetic field from the other coil generates an electrical current in this coil.

If you run an alternating current through one of the coils, it creates a changing magnetic field in and around the coil.[7] As explained earlier in the book, a changing magnetic field produces an electric field. The electrons in the second coil see this electric field and feel a force. In other words, by running an AC current through the first coil, we cause a current to flow in the second coil (Figure 7.4).

---

[7] Just running a direct current through the coil will create a magnetic field, as we learned in Chapter 4. An alternating current means the current is changing constantly, causing the associated magnetic field to change constantly.

To make the transfer more efficient, you can wrap the coils around an iron core, as shown in Figure 7.5. Remember from Chapter 4 that by getting the magnetic moments in iron to line up with an external magnetic field, you can make that field stronger.

Figure 7.5

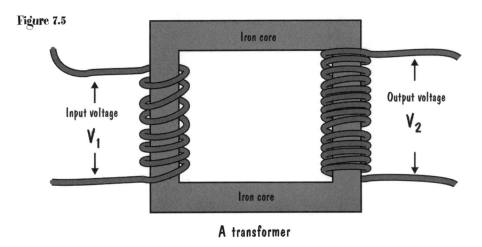

A transformer

Being the observant person you are, you undoubtedly have noticed that the coil on the left in Figure 7.5 has fewer turns than the coil on the right. There's a reason for that. The ratio of the number of turns in one coil and the number of turns in the other coil determines the relative voltage across the coils, according to the following:

$$\frac{\text{number of turns in coil 1}}{\text{number of turns in coil 2}} = \frac{\text{voltage across coil 1}}{\text{voltage across coil 2}}$$

I won't go into the details of how we know this, other than to say that it involves setting the electric power in one coil equal to the electric power in the other coil. If you're familiar with energy concepts,[8] this is equivalent to the expression *work in = work out* for a system. At any rate, what this means is that to transform from low voltage to high voltage (called a step-up transformer) or from high voltage to low voltage (called a step-down transformer), you simply need a transformer that contains just the right ratio of turns of wire in the two coils. And that's what we do. People generate electricity at a low voltage, step it up to a high voltage so it doesn't lose as much energy in transmission, and step

---

[8] This is where I once again take the opportunity to shamelessly promote the *Stop Faking It!* book on Energy!

Figure 7.6

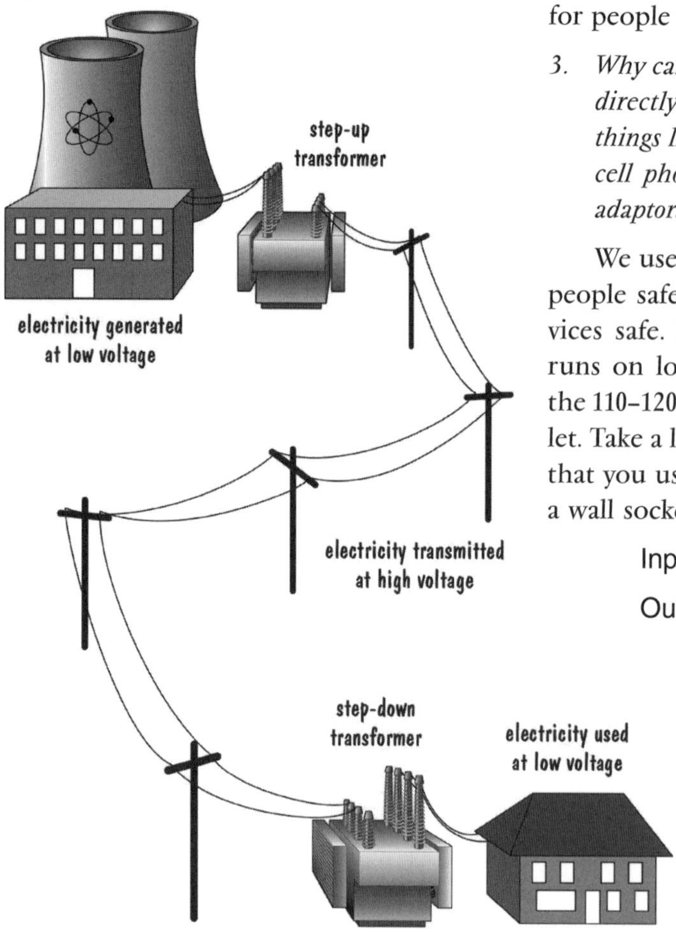

step-up
transformer

electricity generated
at low voltage

electricity transmitted
at high voltage

step-down
transformer

electricity used
at low voltage

it back down to a low voltage so it's safe for people to use (see Figure 7.6).

3. *Why can we plug some things like lamps directly into wall sockets, while other things like computers and train sets and cell phones and electric razors require adaptors?*

We use transformers not just to keep people safe, but also to keep various devices safe. For example, your computer runs on low voltages—much lower than the 110–120 volts available from a wall outlet. Take a look at the special transformer that you use to plug your computer into a wall socket. Mine says the following:

Input: AC 100-240 V Max 1.5 A

Output: 19 V 3.68 A

The V stands for volts and the A stands for amps. Notice that as we reduce the voltage, the current increases, just as with transformers involving high-voltage power lines.

Many transformers, while altering voltage, also transform alternating current into direct current, depending on the device you want to operate. That transformation is relatively easy to accomplish. Keep in mind, though, that transformers used to change voltage *must* use alternating current. Without alternating current, our two-coil setup doesn't get us anywhere. So, the best reason I can think of for you having alternating current available in wall sockets is that we need it in order for transformers to work.

More things to do before you read more science stuff

By the end of this chapter, I'd like you to have at least an idea of one of our most important uses of electricity—the computer that you use to run the software to simulate electric circuits! In order to know how computers work, you need to

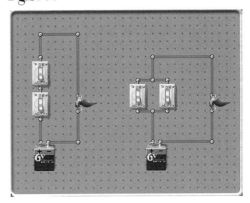

Figure 7.7

Topic: Computers

Go to: *www.scilinks.org*

Code: SFEM27

understand something called **logic circuits**, so let's look at a few. Figure 7.7 has two circuits. Go ahead and build these using the software.

After you've built the circuits and played with them, answer the following question for each circuit. In what position(s) of the two switches will the light be on and in what position(s) of the two switches will the light be off? To get you started, I'll tell you that in the circuit on the left, with switch one on and switch two off, the lightbulb is off; with switch one off and switch two on, the lightbulb is off. Get the idea of what I'm after?

Next build the circuit pictured in Figure 7.8.

What happens to the lightbulb when the wall switch (not the relay switch) is on? What happens to the lightbulb when the wall switch is off? Easy questions, no? If you're interested in what the relay switch is doing when you turn the wall switch on and off, use the schematic view.

One more circuit for you to construct, and then I'll tell you why in the heck I'm having you investigate these silly collections of batteries, switches, and bulbs. That circuit is shown in Figure 7.9.

Let's call the two wall switches switch 1 and switch 2. Our object of interest is the lightbulb at the very top of the circuit. The lightbulb on the lower left is just there to help you realize when current is flowing through that particular wire, and when it's not. Of course, you could also use the "view charges" button to get that information.

Figure 7.8

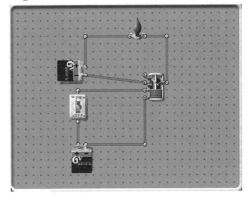

Figure 7.9

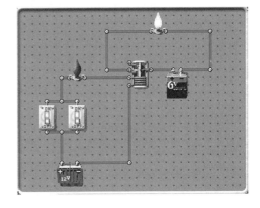

Your task, which isn't all that tough, is to determine what positions of the two wall switches lead to the top bulb being lit, and what positions lead to the top bulb not being lit. If you're trying to think ahead of me, notice that the lower part of the circuit pictured in Figure 7.9 is pretty much the same as the circuit on the right in Figure 7.7, and the top part of the circuit pictured in Figure 7.9 is pretty much the same as the circuit in Figure 7.8.

More science stuff

I already told you that the circuits you just built are called logic circuits, so it's probably best that we discuss what logic is. To do that, put yourself in the following situation. Suppose you have a hankering for cappuccino flavored ice cream, but you have none in the house. You know the grocery store sometimes stocks cappuccino ice cream and sometimes not. The store also isn't open 24 hours a day, so if you're going to be able to satisfy your craving, two conditions must be met: the store must have cappuccino ice cream in stock *and* the store must be open.[9] To be more formal about it,

You can have cappuccino ice cream if the store has it in stock and *the store is open.*

That statement is known as a **logical "and"** statement. The only way the final goal can be achieved is if *both* of the stated conditions are true. The circuit on the left in Figure 7.7 actually represents this logical "and" statement. If the lightbulb is lit, you can satisfy your craving for cappuccino ice cream. One of the wall switches represents whether the store is closed or open. If the store is open, the switch is on and if the store is closed the switch is off. The other wall switch represents whether or not cappuccino ice cream is in stock at the store. If it's in stock, the switch is on and if it's not in stock, the switch is off. Just as both of our real-life conditions must be met in order for you to have your ice cream, both switches must be on in order for the light to be lit.

Now suppose you aren't all that picky, and you'll settle for pistachio almond ice cream instead of cappuccino. In this case, we can make a different kind of formal logical statement, namely

If you have cappuccino ice cream or *pistachio almond ice cream, you are happy.*

Notice that the connecting word here is "or" rather than "and." Not surprisingly, the statement in italics is called a **logical "or"** statement. Also not surprisingly, the circuit on the right in Figure 7.7 models a logical "or" statement (why else would I have you build it?). Each wall switch represents the

---

[9] We could easily impose other conditions on your being able to have cappuccino ice cream, such as the requirement that you have enough money and the requirement that you are not in a coma, but we'll just keep it simple for now and assume all other necessary conditions are met.

presence or lack of each kind of ice cream. The light represents your happiness. If the light is lit, you're happy, and if the light is off, you're unhappy.[10] If both wall switches are off, you have neither kind of ice cream and you're unhappy (the light is off). If either one or both of the wall switches is on, you have ice cream and you're happy (the light is on).

We can take this a step further and combine the two logic statements as well as the two circuits. Whether or not you can get either kind of ice cream is modeled by the left circuit in Figure 7.7—the store must be open and the particular kind must be in stock. Whether you are happy is modeled by the right circuit—either you have one kind of ice cream or the other, both kinds, or neither kind. In the last case you are unhappy. The statement that represents all this is below, and the circuit that represents the statement is shown in Figure 7.10.

Figure 7.10

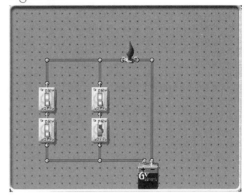

You will be happy if (The store is open and *they have cappuccino ice cream in stock)* or *(The store is open* and *they have pistachio ice cream in stock).*

In the circuit, the light represents your happiness. If the light is lit, you're happy. The two wall switches on the left represent the availability of cappuccino ice cream (the store is open *and* they have cappuccino in stock means both switches are on) and the two wall switches on the right represent the availability of pistachio almond ice cream (the store is open *and* they have pistachio almond in stock).

So far, we know how to create *and* circuits and *or* circuits, and how to combine them. I know you're wondering what the point is, but just stick with me as we look at two more logic circuits, namely the ones in Figures 7.8 and 7.9. What we have in Figure 7.8 is a *not* circuit. When the wall switch is on, the lightbulb is off, and when the wall switch is off, the lightbulb is on. In other words, the top part of the circuit negates whatever happens in the bottom part of the circuit. This is made possible by the way the wires are hooked up to the relay switch. There is an electric current in the top part of the circuit whenever there is *no* current in the bottom part of the circuit. If you missed that, build the circuit again and use the "View Charges" as well as the "View Schematic" buttons to see when current is flowing and when it's not.

Figure 7.9 is similar to Figure 7.8 in that the top part of the circuit negates what happens in the bottom part of the circuit. The bottom part of Figure 7.9 is an *or*

---

[10] Yes, I am making you out to be quite the shallow person if your happiness depends totally on ice cream, but we've all had days like that, no?

circuit. Since the top bulb lights only when no current is flowing in the lower circuit (both switches are off), this circuit is called a *nor* circuit. It gets its name from the fact that it negates an *or* circuit. The top bulb lights if either one or both of the lower switches is on—not! Again, you might want to rebuild this circuit and use the "View Charges" and "View Schematic" buttons to see what's happening.

Now you're *really* wondering why in the world we're messing with these logic circuits. Is it just for fun, or is there a purpose? I'm all in favor of fun, but there definitely is a purpose. It turns out that just about all of mathematics is based on logic, from addition and subtraction to calculus. In the next couple of sections you'll see how we can use logic circuits to add numbers. In the meantime, I'll tell you what's inside your computer, your stereo, your TV, and all other kinds of electrical devices that use logic in order to operate. There are small things called **integrated circuits** that are nothing more than a collection of logic circuits. These tiny circuits are called *gates*, because there are generally two input currents or voltages, with one output current or voltage. Each gate has a name that tells what function the gate performs. There are "and gates," "or gates," "nand gates," "nor gates," and many other kinds of gates. I know that providing names for the gates and showing what the circuits look like doesn't really tell you how all this applies to math. Fortunately, we have another two sections in which I'll help with that.

Even more things to do before you read even more science stuff

Before you start doing things, I'm going to provide a quick explanation of the **binary, or base 2, number system**. If you don't need that review, skip the next few paragraphs. Since you're still reading, I'll assume that binary numbers aren't your strong point. To begin, consider a regular old number like 427. You learned in first grade that this number means 4 hundreds plus 2 tens plus 7 ones. In the base 10 system, which is what we use everyday, there's a ones place, a tens place, a hundreds place, a thousands place, etc., for every number you can

Figure 7.11

think of (we'll ignore decimal places for now). The largest digit you can have in any single place is 9, because adding one more to 9 means you make a mark in the next place over.

Base 10 isn't the only possible number system. You can write numbers in base 4, base 7, base 9, or whatever. For example, in base 7, there is a ones place, a sevens place, a forty-nines place, etc. In base 2, you have a ones place, a twos place, a fours place, an eights place, etc., as in Figure 7.11.

In base 2, all numbers consist of ones and zeros, because once you have counted to one in each column, the next number carries over to the column to the left. For example look at the fours place. If you put a one in that place, it stands for one four. If you were to put a two in that place, it would stand for two fours. However, the column to the left of the fours place is the eights place, so instead of putting a two in the fours place, you should simply put a one in the eights place.

Figure 7.12

Let's take a look at a number written in base 2, otherwise known as a binary number. 10110 is the number 22, because we have a 1 in the sixteens place, a 1 in the fours place, and a 1 in the twos place. This adds up to the number 22, as shown in Figure 7.12.

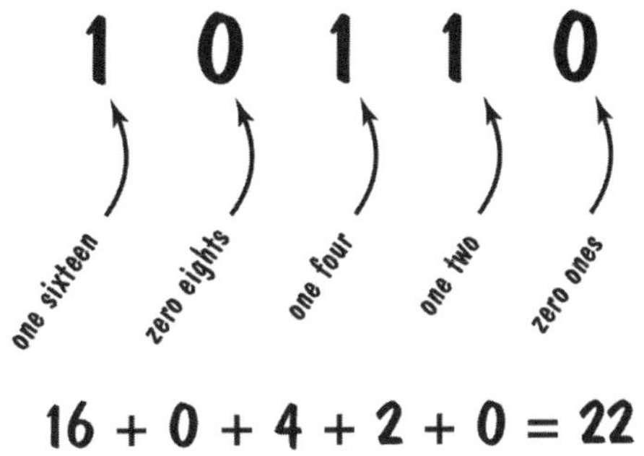

Before messing around with adding binary numbers, I want to point out why we're even bothering with them. Every binary number consists of ones and zeros, which makes binary numbers easy to represent using logic circuits. The result of a logic circuit, no matter how complicated, is that the top light is either on or off. This can stand for there being either a one or a zero in a given number place. Because of this close connection between binary numbers and logic circuits, computers rely almost entirely on binary numbers and binary math—ones and zeros. Hey, why do you think the movie *The Matrix* has all those ones and zeros falling down the screen?

Okay, let's add numbers. First add 53 and 88 in base 10, keeping track of what goes in each place (ones, tens, hundreds), when you carry a 1 over to the next column, and *why* you carry that 1. After you've done that, simply add 1 and 1 in base two. If you do it correctly, you'll have to carry

1 0 1 1 0

one sixteen · zero eights · one four · one two · zero ones

16 + 0 + 4 + 2 + 0 = 22

Topic: Computer
Technology

Go to: *www.scilinks.org*

Code: SFEM28

Figure 7.13

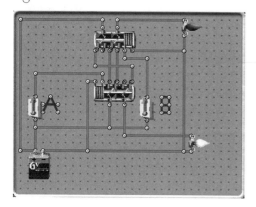

a number from the ones column to the twos column. Finally, build the circuit shown in Figure 7.13 and see if you can discover the connection between this circuit and adding 1 and 1 in base two. Notice that the switches are labeled A and B. Also notice that the two relay switches in this circuit are *double* relay switches, meaning that when the electromagnet is activated, two switches rather than just one are activated.

I realize that you never would have dreamed up this circuit on your own. Just use the "View Charges" button and determine what conditions of the two switches lead to the top light being lit and what conditions of the two switches lead to the bottom light being lit.

Even more science stuff

Let's take a close look at adding the numbers 53 and 88 in base 10.

$$53$$

$$+ 88$$

Starting in the ones place, we're adding 8 and 3. That equals 11, which is a 10 and a 1. Because we have a 10 in that answer, we carry that ten over to the tens place and leave the 1 in the ones place. Now in the tens place we have 8 tens plus 5 tens, plus the extra ten we carried over from the ones place. That's 14 tens. 10 of those 14 tens equal 100, so we carry that over to the hundreds place, leaving 4 tens in the tens place. Our final answer is 141. Easy, huh?

Now let's add 1 and 1 using base two.

$$1$$

$$+ 1$$

We all know that 1 plus 1 equals 2, but in binary arithmetic, we don't use the number 2. Because 1 plus 1 equals 2, we have to carry a 1 over to the twos place, leaving a zero in the ones place. Our answer is therefore 10. In base ten, the number 10 represents the number 10, but in base two, 10 represents the number 2.[11]

---

[11] A joke that I credit to reviewer Dr. Larry Kirkpatrick, who undoubtedly can credit someone else: "There are only 10 kinds of people in the world—those who understand binary numbers and those who don't."

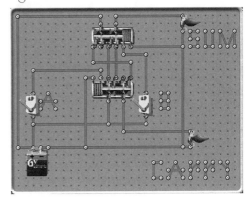

Figure 7.14

If you followed that, then take a look at the circuit in Figure 7.13, which is reproduced in Figure 7.14, with a couple of extra words added.

We're going to use this circuit to add in base two. The wall switches A and B represent the digits in the ones place for each of the two numbers we're adding. If a switch is on, the digit in the ones place is a 1. If the switch is off, the digit in the ones place is a zero. The top light in the circuit is labeled "sum." If this light is on, then there's a 1 in the ones place after adding. If the light is off, there's a 0 in the ones place after adding. The bottom light is labeled "carry." If this light is on, you carry a 1 over to the twos place. If this light is off, you don't carry anything.

Suppose you want to add the binary number 1 to the binary number 0. Then switch A is on and switch B is off. This lights the top bulb, indicating that you need a 1 for the sum in the ones place. The bottom light is off, meaning you don't carry anything over to the twos place. Now add 1 and 1, meaning switches A and B are both on. Notice that this results in the top lightbulb being off, meaning you put a 0 in the ones place in the sum. The bottom light, however, is on, meaning you have to carry a 1 over to the twos place.

By connecting a bunch of circuits like this together, you can add any two binary numbers. You will need a circuit like Figure 7.14 for each place in the sum. Each circuit has a switch for the number (either a 1 or a 0) in that place for the digit in the numbers you're adding, plus another input switch for the possibility that you have to carry over a digit from the previous place.

What I'm describing is a computer. You can hook enough logic circuits together to add, subtract, divide, and multiply any numbers you want. Although the circuits and combinations of circuits get more complicated, you can also use them to do algebra, geometry, and calculus. Because everything a computer does, even when you're not doing math, comes down to logical actions that depend on conditions, you can set up circuits that will allow you to type and print words, play video games, and all sorts of things. Of course, there's a big problem with what we're doing, and the key is the word *big*. If you set up enough electric circuits just to perform simple additions, you can easily have so many components that you fill a room with your computer. In fact, early computers *were* quite large. For an independent project in college I worked on programming a missile guidance computer that was built in the '60s. That thing was

Topic: Semiconductors
and Insulators

Go to: *www.scilinks.org*

Code: SFEM29

gigantic, and it made so much noise you couldn't talk to someone in the same room if it was on. Ah yes, the good ol' days. I mentioned earlier something called *integrated circuits*. These integrated circuits rely on components known as **semiconductors** to produce really, really tiny logic circuits. Those circuits are so tiny that the computer you have in front of you is many times more powerful than a computer that would have filled a football stadium back in the '70s. Same logic, though.

This is where I'm going to end the book, although we have barely touched the many kinds of things electric circuits can do. That's the nature of the books in this series, though. I want you to understand the basics of electricity, magnetism, and circuits well enough that you can not only teach those concepts with confidence but can tackle more advanced treatments of the subject with the necessary knowledge base. Good luck with that!

I also hope you realize there are hours of playtime fun left in the electric circuit software. The "Ideas" button contains lots of cool circuits to explore, and of course you can try your own combinations without fear of burning up your house. If you ever get stuck in figuring out how a circuit works, remember that the "view charges" button can always help you figure out where electric charges are flowing and where they're not.

Chapter Summary

- Electric current generally comes in two forms—direct current (DC), which travels in a given direction, and alternating current (AC), which changes direction at a given frequency.

- One can generate an alternating current with a generator or turbine and one can generate direct current with a generator or a chemical reaction such as might be found in a battery.

- People generally generate and use electricity at low voltages. In order to reduce heating losses, though, we generally transmit electricity at high voltages.

- Transformers can transform low-voltage, high-current electricity to high-voltage, low-current electricity, and vice versa.

- Electric circuits can model logical statements, which are the basis of many operations performed by computers.

- Computer operations are based on binary mathematics (base two), which is easy to model with logic circuits.

- Despite the power of binary math, I seriously doubt Neo can do some of those climbing-the-wall tricks or dodge bullets with so little effort.

Applications

1. You already know from previous chapters that electric components can get rather hot when they're operating (joule heating). Some devices have what are known as "heat sinks" that do a good job of carrying heat away from components. The heating up of components in computers is an especially important problem. Because so many tiny, tiny electric circuits are jammed into tiny, tiny spaces, the integrated circuits that make up computers are quite delicate. Even moderate increases in temperature can affect the operation of computers. Therefore, your computer has a fan that operates continuously while you're computer is on. If you can't hear the computer's fan while you're working, try restarting your computer. Still no fan? See an official repair-type person. Operating a computer without any means of keeping the components cool is a sure way to be in the market for a new computer.

2. You know that by turning a coil of wire inside a magnetic field you can generate electricity. That's called a generator. Did you also know that by running an electric current through a coil of wire that happens to be immersed in a magnetic field, you can cause the coil of wire to turn? It's a basic application of the interaction of moving charges and magnetic fields. You have a coil of wire that's sitting in the field of some external magnet. When you run an electric current through the wire, that means you have moving charges in that wire. We also know that moving charges that are immersed in a magnetic field feel a force. If you set things up just right, the forces on those moving charges can cause the coil of wire to spin. I bet if you're really clever, you can hook that spinning coil of wire up to a wheel or two and cause something to move. What I've just described is a **motor**. A motor is the same device as a generator—just one that's set up to operate in an opposite sense. In a generator, you turn a coil of wire immersed in a magnetic field to generate electricity. In a motor, you run an electric current through a coil of wire immersed in a magnetic field to cause the coil to spin.

Glossary

alternating current (AC). An electric current that changes direction rapidly. The current available from regular wall outlets is alternating current.

amperes. The units in which electric current is measured. Most likely the source of the term *amped,* which you'll hear among snowboarders and skateboarders.

atom. A very tiny thing that we can't see directly. Part of a scientific model for what matter is made of.

attractive force. A force between two objects that pulls them together. Also, a well-dressed force.

bimetallic strip. A metal strip that is actually composed of two different kinds of metals. A bimetallic strip bends or straightens depending on its temperature, and is the central component of a thermal switch.

binary number system. A number system based on the number 2 rather than the number 10. Computers rely on the binary number system for basically everything they do, mainly because it's easy to model binary math with electric circuits.

capacitor. An electric device used to store charge in circuits. If you go poking around in unplugged high-voltage equipment with a screwdriver, you just might find a large capacitor that is still charged up and you just might find yourself thrown against the wall by the jolt. Happened to a fellow graduate student!

circuit breaker. A device that opens an electric circuit when the current in the circuit exceeds a certain value. Circuit breakers protect devices in electric circuits and prevent fires, so they're good things.

complete circuit. An electric circuit that has no gaps in it and makes a continuous path, allowing current to flow through the circuit. Also, a circuit that has recently found love.

conduction. The process of positive or negative charges moving from one place to another. In reference to an electroscope, charging by conduction means touching a charged object to the electroscope.

conductor. A material that allows electric currents to flow through it without much difficulty. Also someone who hangs around trains and looks important in a fancy uniform.

coulombs. The units in which one measures the quantity of electric charge.

Coulomb's law. A mathematical relationship that expresses the strength of the force between two charged objects.

current electricity. The name given to charges that are in motion.

diamagnetic. The description given to a material whose atom's magnetic moments are either very small or tend not to line up with one another, or both.

direct current (DC). An electric current that flows in only one direction. Batteries produce direct current.

electric circuit. The name given to any collection of wires and/or electric components.

electric dipole. Any arrangement of charges that has a definite separation of positive and negative charges. Many atoms and molecules are naturally occurring electric dipoles.

electric field. A scientific model that one can use to explain how electric charges affect one another.

electric monopole. Any isolated positive or negative charge.

electrically neutral. The description of anything that contains an equal number of positive and negative charges. Objects in the universe tend toward being electrically neutral. I'm sure the Swiss people approve.

electromotive force (emf). The generic name given to any device that causes an electric current to flow continuously.

electromagnetism. The name given to the combined subjects of electricity and magnetism. It's a good name, because electricity and magnetism are different aspects of the same phenomenon.

electron. A negatively charged particle usually found hanging around the nuclei of atoms. Electrons are so small that physicists treat them as "point objects," meaning they have no size at all!

electron cloud. A description of how electrons fit into our model of the atom. They can best be thought of as residing in a "probability cloud" around the nucleus. Not the most satisfying picture, but certainly more accurate than one of electrons moving in orbits around the nucleus. Also, something rock star electrons are heard telling other electrons to get off of (yes, that is an obscure reference!).

electrophorus. A device you can use to charge up other things over and over without having to do a whole bunch of rubbing each time.

electroscope. A device you can use to determine the sign of the charge on an object as well as the magnitude of the charge on the object.

exchange coupling. A quantum mechanical effect that causes the magnetic moments of electrons in a material to act as one. I didn't really explain exchange coupling in this book, but it is the basis for viewing magnetism as being due to the growing and/or shrinking of magnetic domains in a material. So yeah, you can ignore this term if you want.

ferromagnetic. The description given to a material whose magnetic moments line up easily and then tend to remain lined up. Permanent magnets are composed of ferromagnetic materials

frequency. A number that represents how often an alternating current changes direction. The frequency of the alternating current available from a wall outlet is usually 60 cycles per second.

fuse. An electric device that fries when the current through it goes above a certain magnitude, thus creating an open circuit and preventing current from flowing. Fuses protect electric devices and prevent fires, but many fuses are now replaced by circuit breakers. Also, something that can be short on any family member at any given time.

galvanometer. A meter that indicates the magnitude of an electric current. Its operation depends on the fact that moving charges (an electric current) feel a force when immersed in a magnetic field.

generator. A device that can transform the energy of moving wires or magnets into an electric current.

grounding. The process of connecting something to an essentially infinite reservoir or source of excess charge. That something is usually the Earth. Also, something parents do to teenagers and later regret because they are then stuck with the problem 24 hours a day.

impedance. The concept when dealing with alternating current that is equivalent to the concept of resistance when dealing with direct current.

induced charge. The separation of charges in a neutral object so the object acts like a charged object rather than a neutral object.

induced magnetism. The process in which magnetic moments of a material become lined up due to the influence of an external magnetic field. This lining up causes the material itself to act like a magnet.

induction. In reference to an electroscope, this describes the process of charging the electroscope without actually touching the electroscope with a charged object.

insulator. A material that does not generally conduct an electric current. Common insulators are rubber, air, and wood.

integrated circuit. A collection of various logic circuits put together in a rather small package. Integrated circuits are essential components of most modern electric devices, from televisions to computers.

joule heating. The term applied to heat energy that is lost when an electric current flows through any material. Also, a great name for a furnace company that specializes in catering to physics nerds.

Kirchhoff's rules. A couple of rules that describe what happens in electric circuits. One is the junction theorem, which states that the electric current entering any circuit junction is equal to the electric current leaving that junction. The other is the loop theorem, which states that the sum of all voltage rises and drops around any loop in an electric circuit is equal to zero.

logic circuits. Electric circuits that operate using the formalism of logic statements.

logical "and" statement. A statement that says C is true only if *both* A and B are true. Those letters A, B, and C can represent various conditions in all sorts of real-world situations, or they can represent purely mathematical conditions.

logical "or" statement. A statement that says C is true if *either* A or B is true. A, B, and C can represent various conditions in all sorts of real-world situations, or they can represent purely mathematical conditions.

magnetic field. A scientific model that one can use to explain how magnets affect one another and how they affect moving charges

magnetic dipole. Any configuration of magnets, or things that act like magnets, that has a separated north and south magnetic pole. Basically, whenever you deal with magnets, you are dealing with magnetic dipoles.

magnetic domains. A scientific model in which magnetism, or lack thereof, is explained in terms of regions of magnetism in a substance that grow and diminish depending on external influences.

magnetic moment. A concept that represents individual atoms or molecules as behaving like tiny magnets. Also, what happens when you finally meet that special person.

magnetic monopole. Something that, to our present knowledge, does not exist, If it existed, it would be a completely isolated north or south magnetic pole.

magnetic poles. A scientific model in which magnets of all sorts contain two kinds of poles, north and south. Like poles repel and unlike poles attract.

motor. A device that takes an input of electric current and transforms it into some kind of mechanical motion, usually a turning motion. The operation of motors is based on the concept that moving charges immersed in a magnetic field feel a force.

neutron. A particle that has no charge and is contained in most nuclei of atoms. Also, the last name of a boy genius who has a really funny friend named Carl. Only parents will get that last reference.

nucleus. The name for the center part of an atom. The nucleus consists of protons and neutrons, and a slew of other tiny particles that you would only know about if you studied recent theories of subatomic physics.

ohms. The units of resistance.

Ohm's law. A mathematical relationship between the voltage across an object, the current through the object, and the resistance of the object. It's written as $V = IR$. Many materials obey Ohm's law for all intents and purposes, but some don't obey the law at all, and others only obey the law within certain voltage and current ranges.

parallel circuit. A circuit, or even a small section of a circuit, in which electric current has more than one path it can take.

paramagnetic. The description given to a material whose magnetic moments tend to line up with an external magnetic field, but which then tend to return to a random orientation.

potential difference. The name given to a region in an electric circuit in which charged particles would tend to move from one place to another. When using a gravitational analogy, potential difference is a difference in height. The greater the physical height in the analogy, the greater the potential difference in the electric circuit.

power. The amount of energy produced or consumed per unit time. Also, something all politicians lust after.

probability cloud. A model that best explains how electrons are located in an atom.

proton. A positively charged particle that generally resides in the nucleus of atoms. As with many scientific models, you will never be able to see a proton directly. You will only be able to observe effects that make sense if we assume the presence of protons in the universe. Yes, it's okay if you don't believe in protons. They are part of a model, not a truth that is evident to anyone with a brain.

relay switch. A switch used in electric circuits that redirects the path of electric current depending on the current inputs to the switch.

repulsive force. A force between two objects that pushes them apart.

resistance. A measure of how much a given device restricts the flow of electric current. Resistance is measured in ohms.

resistor. A device one uses in electric circuits to slow or alter the flow of electric current. Also, someone who doesn't agree with the political status quo.

semiconductor. A material that conducts electric current only under certain conditions. Semiconductors are the basis for most of the electronic devices we use today.

series circuit. A circuit, or even a small section of a circuit, in which electric current has only one possible path to follow.

static electricity. The name given to interactions among charges that are stationary.

thermal switch. An electric switch that turns on and off depending on how much current is flowing through it. Thermal switches usually rely on bimetallic strips.

transformer. An electric component that can transform the voltage and current of an alternating current. Transformers are responsible for converting the high voltage electricity that heads toward your home into low voltage electricity that will generally give you a bit jolt if you misuse it, but won't kill you.

Triboelectric series. A list of substances ordered according to their tendency to attract or give up electrons.

vector. A representation of quantities that have both magnitude and direction. A vector is represented by an arrow, with the direction of the arrow indicating the direction of the quantity and the size of the arrow indicating the magnitude.

volts. The units of potential difference.

watts. The units of power. One watt is equal to one joule per second.

Index

*Page numbers in **boldface** type refer to figures.*

R

Relay switch, 114–115, 133, 148

Repulsive force, 4, 38–41, 148

Resistance, 86–92, **87–91,** 148
of human skin, 94
impedance and, 126
of wires, 99

Resistors, 88, 92, 98–99, 149
adding in parallel, **98,** 102–104, 111–112, 119
combining in series, **98, 101,** 101–102, 119
transmission lines as, 129

Right-hand rule, 63, 73

S

Safety precautions, 52, 93–94, 121

Scientific explanations, 3

SciLinks, x
batteries, 106, 118
computer technology, 137
computers, 133
current electricity, 92
Earth's magnetic field, 46
electric circuits, 78
electric current, 55, 81
electric power, 96
electrical energy, 22, 128
electrical safety, 52
electromagnetism, 61
electromagnets, 74
electron, 5
electronic circuits, 86
generators, 71
insulators, 93
magnetic fields, 45
magnetic fields of power lines, 128
magnetic materials, 40
properties of magnets, 42
semiconductors and insulators, 140
static electricity, 6
types of magnets, 38

"Sea floor spreading," 73, **73**